本书编委会 \ 编

域 下

中国室内设计年鉴

INTERIOR DESIGN YEARBOOK OF CHINA

中国林业出版社
China Forestry Publishing House

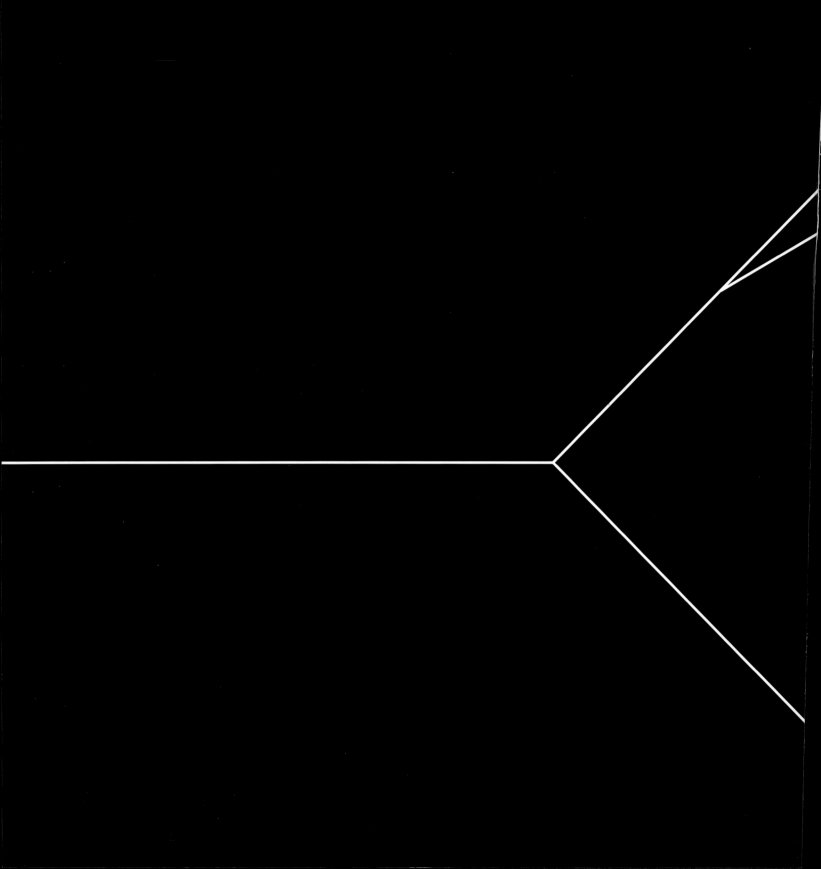

设计版图!

近年来,中国的室内设计取得了长足的进步,本土的设计力量逐渐浮出水面,这让我们看到了中国设计崛起的可能性,但某种程度上,境外的设计师力量仍然占有某些优势。中国是全世界范围内设计最为蓬勃发展的设计热土,每年这里都会涌现出无数的作品,这催生了设计出版行业的繁荣,这些图书及杂志为我们记录了当代中国设计的发展动态,是我们了解当下中国设计的重要路径。坊间每年都会出版几本中国室内的年鉴类书籍,他们是我们了解本年度室内设计发展的重要窗口,但是这些室内年鉴在编选的时候,侧重于本土设计力量,而且是按照项目类型来编辑的。因此,我们尝试以一种新的方式来解读当下中国设计的新发展。我们没有按照项目的类型来编辑设计作品,而是按照行政区域来划分,而且我们的视野不仅停留在本土的设计力量,而是关注一切在中国的设计力量。

中国幅员辽阔,地区间社会、文化及经济发展存在非常大的差异,相应地,设计的发展状态也不太一样,尝试去勾勒一个大致的中国当下设计版图也许有助于我们更加理性认识当下中国设计的发展及状态。大致来说,目前大陆地区,设计最为活跃、最为繁荣的区域是华南地区、华东地区、华北地区,华南地区以深圳、广州等为代表,华东地区以上海、苏州、杭州、厦门、福州等地为代表,而华北地区则以北京、天津为代表。当然,勾勒中国的设计版图不可能少了香港、台湾。这些区域涌现出的设计作品构成了我们本书的主体。这些区域之外,华中地区、东北地区也有一些优秀的设计作品呈现,但是西北地区及西南地区的设计作品较少进入我们的视野。一方面,这也许受我们自身视野所限,但另一方面也显示了目前西南及西北区域的设计力量仍然比较薄弱。

虽然我们尽力勾勒当下中国清晰的设计版图,但这异常困难。当下中国每年涌现的优秀作品实在不胜其数,由于能力的限制,我们在编选的时候,难免会遗落大量的杰作。但是我们希望读者可以从我们选择的作品中,看到本年度中国当代室内设计的发展动态,窥一斑而见全豹。上海、北京、香港、台湾、深圳等地是传统的设计活跃城市,在这些地方不仅仅本土设计力量强劲,而且国际性的设计力量也非常强大。在华东区域,厦门、福州是非常值得一提的城市,这两个城市的设计力量近年来非常活跃,在国际国内都获得了非常多的奖项,他们浮出水面是中国设计取得长足进步的结果。华北区域的天津虽然也非常值得关注,近年来也涌现了非常多的优秀作品,但这些作品的设计师不是本土的设计力量,而是来自境外或者上海、北京、深圳等地设计力量。这显示当地设计力量的发展得益于政策的推动,但仅仅只有政策的推动并不足以让当地的设计力量成长,还需要当地设计力量自我学习及反思。西南区域的重庆,近来年也成为了各种设计力量角逐的战场,据说当地每年要兴建十几个五星级酒店,由此可见当地设计市场的火爆,但是如此火爆的设计市场并没有孕育出对称的设计作品。这其中的缘由值得我们反思。香港设计对内地特别是深圳设计的影响早已成为共识,但是一直以来,我们虽然也非常关注台湾的设计力量及其对大陆设计的影响,但我认为,我们的关注仍然不足。台湾地区的设计既有中国传统文化的影响但又非常国际化,在我看来,这是中国当代设计自成一格的不二法门。因此,在本书中,我们特别将台湾地域辟为一章,以彰显其重要性,也希望引起业内对台湾设计的进一步认识。

In recent years, China whose local design strength has been revealed has made considerable progress in interior design, which makes us see the possibility of rise of Chinese design, but to some extent, oversea designers force still occupy some advantages. China is the most booming design hot spot worldwide and numerous works emerge every year here, which has given birth to prosperity of design and publishing industry recording development trend of contemporary Chinese design by books and magazines for us as the important path that we learn about contemporary Chinese design. Several indoor Yearbooks newly published every year are known as an important window to understand interior design development of the current year, but these indoor Yearbooks when compiled have been focused on local design strength and edited in accordance with the project type. Therefore, we try a new way to interpret the new development of contemporary Chinese design. Instead of project type, we have used administrative region division to edit design works, and in addition, we focus our vision not only merely on the local design force but also on all design strengths in China.

Many social, cultural and economic development differences between regions of China as a large country, accordingly result in various development design states, so that how to outline a general contemporary design layout in China may help us to more rationally understand development and state of contemporary Chinese design. Generally speaking, the most active and most prosperous design regions in the mainland are South China represented by Shenzhen, Guangzhou, etc, East China represented by Shanghai, Suzhou, Hangzhou, Xiamen, Fuzhou, etc, North China represented by Beijing, Tianjin, etc. Of course, it is impossible to outline China's design layout without Hong Kong and Taiwan whose design works emerged to constitute the main body of our book. Besides the above mentioned areas, Central China and Northeast also present some excellent design work, but the design work from Northwest and Southwest areas is beyond our field of vision. On the one hand, our own vision may be limited, but on the other hand, also it shows that design strength of Southwest and Northwest regions remains relatively weak.

Although we make every effort to outline Chinese current clear design layout, it is extremely difficult. At the moment, China's outstanding works annually emerge is really numerous, due to capacity constraints, during compilation, we could inevitably neglected a lot of masterpieces. But we wish our readers to see the development state of Chinese contemporary interior design of the current year from 100 pieces of works that we have chosen, jumping into conclusion through fragment observation. Traditionally, design activities are vibrant in such cities as Shanghai, Beijing, Hong Kong, Taiwan, Shenzhen, etc, whose not only local but also international design strength is very powerful. In the region of East China, Xiamen, Fuzhou is very worth mentioning, whose design strength are very active in recent years reputed by lot of international and domestic awards, resulting from great progress in Chinese design. Tianjin of North China is also very worthy of attention, also emerging with a lot of good works in recent years whose designers are from design strength of oversea or Shanghai, Beijing, Shenzhen and other places, but not local. This shows that the development of local design strength benefits from policy promotion, but the growth of local design strength can not be driven only by policy promotion and local design strength needs self-learning and profound consideration. Chongqing of Southwest China where more than a dozen of local five-star hotels are built up every year has become the battlefield for various design strengths during recent years, which indicates how hot its local design market is, but such a popular design market fail to breed symmetrical design works. The reason is worthy of our profound consideration. Comparing with common view over the impact by Hong Kong design on the Mainland's design, in particular on Shenzhen design, we also focus much on impact by Taiwan design strength on the Mainland's design, but we believe our concern is still not enough. Taiwan design, not only impacted by traditional Chinese culture but also very international, in my opinion, is the sui generis one and only way of Chinese contemporary design. Therefore, in this book, we particularly set up one chapter for Taiwan design so as to highlight its importance, and also hope to attract the further understanding of Taiwan design industry.

Design layout!

目录 A

002 序 INTRODUCTION

设计版图！/Design layout!

010 华东 REGION EASTERN CHINA

012 和平饭店/Peace Hotel

022 华尔道夫酒店/Waldorf Astoria Hotel

032 上海玻璃博物馆/Shanghai Museum of Glass

042 上海外滩英迪格酒店/Hotel Indigo Shanghai on the Bund

054 and… - SuperPress - SuperBla办公室/
SuperPress SuperBla Office

060 隐泉之语/Haiku Sushi

066 Kartel 酒吧/Kartel Wine Lounge

072 半岛1919红坊艺术设计中心/
Designed and transformed by Bandao 1919 Red Workshop Art Design Center

080 雷迪有限公司办公室/The office of Leidi Limited

088 PARK 97 钢琴吧/PARK 97 Piano Bar

092 五维茶室/Tea House

098 上海卓美亚喜玛拉雅酒店/
Shanghai Zhuomeiya Himalayas Hotel

106 华山路 Chowhaus 餐厅/Chowhaus at Huashan Rd.

114 烧肉达人日式烧肉店/
YAKINIKU MASTER Japanese barbecue restaurant

120 上海爱莎金煦全套房酒店/
Golden Tulip Ashar Suites Shanghai Central

130 南京金地名京/Nanjing Golden Palace

136 南京紫轩餐饮会所/Nanjing Zixuan Restaurant Club

140 江苏亚明室内建筑设计有限公司办公室/
Jiangsu Yaming Interior Architecture Office

146 瓦库6号/Waku No.6

154 万豪华府会所/Wanhao Huafu Club

160	星光捌号 / "Starry Night Dining"	246	皇室戏院 / Boutique Cinema
168	无锡灵山精舍 / Wuxi Lingshan Jingshe	252	设计集人有限公司办公室 / The Office of Design Systems Ltd
174	扬州富临壶园府邸 / Yangzhou Fulin Hu Garden	258	Azura / Azura
180	裸心谷 / Pure Heart Valley		
192	万科良渚文化村少儿会所二期 The second phase of Vanke Liangzhu Culture Village Children Club	**266**	台湾 REGION TAIWAN
196	恒立布业销售中心 / Hengli Cloth Industry Sales Center	268	台北W饭店 / Taipei W Hotel
202	UTI品牌女装文一店 / Design Concept	282	大城朗云接待中心 / Dacheng Langyun Reception Center
208	上虞宾馆 / Shangyu Hotel	292	草山水美 / House For 2
220	零壹城市事务所办公室 / LYCS Architecture Office	298	西园29服饰创作地 / Fashion Institute Taipei
224	杭州环球MUSE酒吧 / MUSE Global Hangzhou	306	咏真接待中心 / Yongzhen Reception Center
230	嘉兴月河客栈 / Jiaxing Yuhe Inn	314	晶宴中和会馆 / Zhonghe-Amazing Hall
		322	欧德旗舰店 / Oude Flagship Store
		330	早安清境 / In The Morning Forest
238	香港 REGION HONGKONG	338	愿意 / Willing
240	嘉禾黄埔电影院 / GH whampoa Cinema		

目录 B

- 002 序 INTRODUCTION
- 设计版图！ Design layout!

酒店 HOTEL

- 012 和平饭店 Peace Hotel
- 022 华尔道夫酒店 Waldorf Astoria Hotel
- 042 上海外滩英迪格酒店 Hotel Indigo Shanghai on the Bund
- 098 上海卓美亚喜玛拉雅酒店 Shanghai Zhuomeiya Himalayas Hotel
- 120 上海爱莎金煦全套房酒店 Golden Tulip Ashar Suites Shanghai Central
- 168 无锡灵山精舍 Wuxi Lingshan Jingshe
- 180 裸心谷 Pure Heart Valley
- 208 上虞宾馆 Shangyu Hotel
- 230 嘉兴月河客栈 Jiaxing Yuhe Inn
- 268 台北W饭店 Taipei W Hotel

办公 OFFICE

- 054 and… - SuperPress - SuperBla办公室 SuperPress SuperBla Office
- 080 雷迪有限公司办公室 The office of Leidi Limited
- 140 江苏亚明室内建筑设计有限公司办公室 Jiangsu Yaming Interior Architecture Office
- 196 恒立布业销售中心 Hengli Cloth Industry Sales Center
- 220 零壹城市事务所办公室 LYCS Architecture Office
- 252 设计集人有限公司办公室 The Office of Design Systems Ltd
- 298 西园29服饰创作地 Fashion Institute Taipei

住宅 HOUSE

- 130 南京金地名京 Nanjing Golden Palace
- 258 Azura Azura
- 292 草山水美 House For 2

餐厅 RESTAURANT

- 060 隐泉之语 Haiku Sushi
- 106 华山路Chowhaus餐厅 Chowhaus at Huashan Rd.
- 114 烧肉达人日式烧肉店 YAKINIKU MASTER Japanese barbecue restaurant

136 南京紫轩餐饮会所 / Nanjing Zixuan Restaurant Club	文教医疗空间 Culture and Education for Medical
160 星光捌号 "Starry Night Dining"	032 上海玻璃博物馆 Shanghai Museum of Glass
174 扬州富临壶园府邸 Yangzhou Fulin Hu Garden	072 半岛1919红坊艺术设计中心 Designed and transformed by Bandao 1919 Red Workshop Art Design Center
314 晶宴中和会馆 Zhonghe-Amazing Hall	192 万科良渚文化村少儿会所二期 The second phase of Vanke Liangzhu Culture Village Children Club

茶馆和咖啡厅 TEA & COFFEE SHOP

092 五维茶室 Tea House

146 瓦库6号 Waku No.6

240 嘉禾黄埔电影院 GH whampoa Cinema

246 皇室戏院 Boutique Cinema

商业空间 BUSINESS SPACE

202 UTI品牌女装文一店 Design Concept

322 欧德旗舰店 Oude Flagship Store

338 愿意 Willing

酒吧和会所 BAR & CLUB

066 Kartel酒吧 Kartel Wine Lounge

088 PARK 97钢琴吧 PARK97 Piano Bar

224 杭州环球MUSE酒吧 MUSE Global Hangzhou

售楼处 SALES CENTER

154 万濠华府会所 Wanhao Huafu Club

282 大城朗云接待中心 Dacheng Langyun Reception Center

306 咏真接待中心 Yongzhen Reception Center

330 早安清境 In The Morning Forest

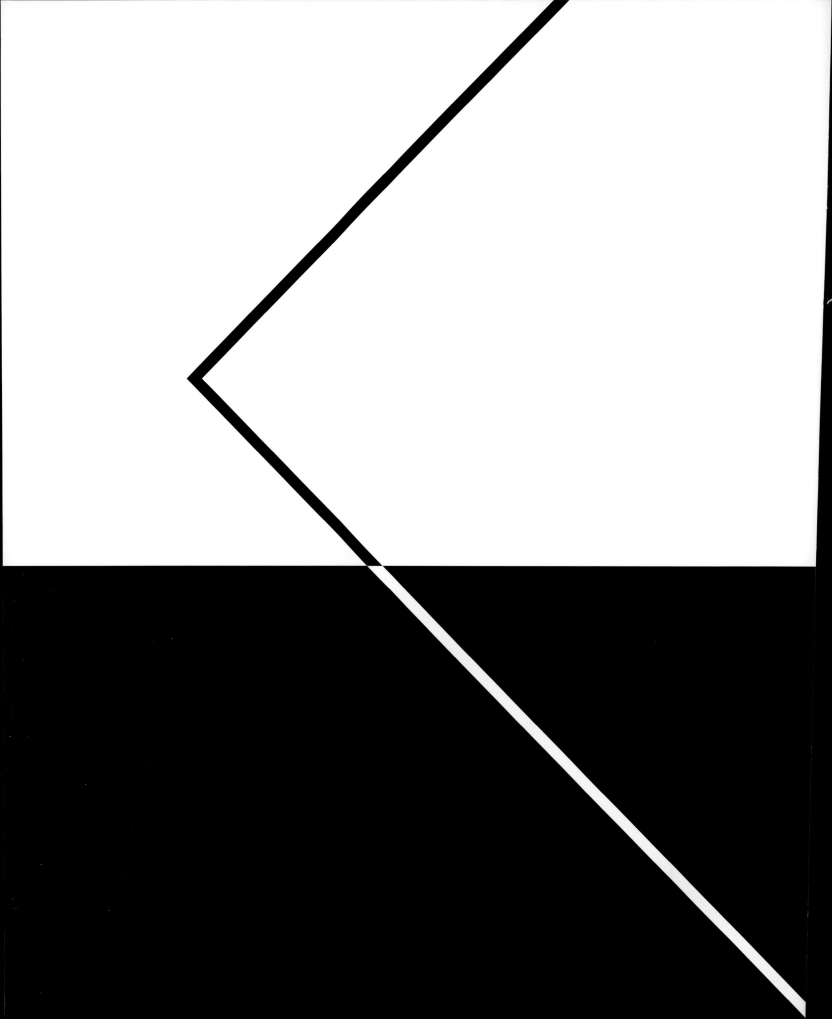

REGION 华
EASTERN 东
CHINA

和平饭店

/项目地点：上海
设计单位：**HBA**

和平饭店位于上海外滩，建筑为装饰艺术（Art Deco）风格。老和平饭店是1920年代和1930年代上海这个"东方明珠"鼎盛时期的标志。当时这座中国最好的酒店号称是"人生极致奢华之地"，曾接待过众多贵客，其中包括喜剧大师查理·卓别林和剧作家诺埃尔·科华德。HBA总监Ian Carr表示："一个多世纪以来和平饭店一直是上海的标志性建筑。她是中国乃至亚洲最有名的饭店。我们力争重现这座亚洲标志性建筑的辉煌和典雅，使之再次雄居世界顶级酒店之列。"

新和平饭店约有256间豪华客房和套房。酒店将精心设置五个餐厅和酒吧，包括位于底楼的让人倍感亲切的爵士酒吧、咖啡厅和大堂吧，位于夹层的寿司吧和美酒雪茄吧，八楼的经典中餐厅和著名的和平扒房。

和平厅也在八楼，那里的舞厅安装着著名的弹簧跳舞木地板，还附设几个会议室以及一个宽阔的室外阳台。酒店后部一座新增的低层建筑将设部分客房，还有一个露天游泳池和水疗房。

著名的"九国特色套房"仍将是新饭店的一大特色：其中四个（印度、英国、中国和美国套房）将保持原貌，而法国、意大利、西班牙、日本和德国套房将在遵循最初理念的前提下进行重新设计。总统套房位于顶层十楼，这里曾是和平饭店的显赫张扬的缔造者和旧主人维克多·沙逊曾经居住的地方。

HBA 的和平饭店设计方案会将再现上海著名的装饰艺术传统，辅以流线型的陈设和现代的室内设备。"这将是典型的 HBA 特点：奢华、现代，又让人倍感亲切；一切浑然天成，为和平饭店这一历史建筑量身定制，"HBA 总监 Connie Puar 这样说。

底楼原来设计的是豪华拱廊商场，现在将恢复原先的古典十字形楼面设计，在宾馆四面均设有旋转门。当年绚烂的八边形玻璃天窗和整个夹层，几十年来一直用石膏板覆盖着，将重见天日。装饰有石子马赛克图案的地板，将呼应原来的装饰艺术风格的瓷砖。

一种柔和的"淡黄泛蓝灰"的色彩方案将使酒店原先精致的上楣柱和天花板增色不少。重新磨光的铜制扶手和轻盈栏杆，由于附加了古铜和抛光的镍而颇具韵味。灰色纹理的大理石四周用瑰丽的法国圣劳芝深黑大理石镶拼，还用核桃木进行装饰，再现了 1930 年代装饰艺术盛行时的风格……各种细节无不真实再现了当年的风貌。

HBA Designs Famous Peace Hotel:
A Global Icon of the 21st Century

The heritage art deco Peace Hotel, located on the riverfront Bund, is being refashioned as an iconic international hotel for the 21st century.

HBA/Hirsch Bedner Associates – the global leader in hospitality design – is now in the final design stages of the world-famous Peace Hotel and transforming it into the ultra-luxurious Fairmont Peace Hotel Shanghai. The reopening of this Shanghai landmark in 2010 will be highly anticipated by discerning travelers worldwide. It will also mark the rebirth of an Asian institution, as the old Peace Hotel came to symbolize Shanghai's Pearl of the Orient heyday in the 1920s and 1930s. Back then, China's finest hotel was known as the "ultimate venue for life's pleasures," and hosted scores of celebrity guests, including Charlie Chaplin and Noel Coward.

The newly revitalized Fairmont Peace Hotel Shanghai will offer approximately 256 deluxe guestrooms and suites. A selection of six restaurants and lounges will include the endeared Jazz Bar, Deli Cafe and a lobby lounge on the ground floor, a mezzanine-level sushi, wine and cigar bar, and a heritage Chinese restaurant and the Peace Grill Restaurant on the eighth floor.

This level will also host the Peace Hall, with its famed sprung-timber dance floor, plus several meeting rooms and an expansive outdoor terrace. A low-rise extension added to the rear of the hotel will house some guestrooms, plus a sky-lit swim-

ming pool and spa.

The famous 'Nine Nations Suites' will remain a feature of the new hotel: four of these (Indian, English, Chinese and American) will be preserved from the old Peace Hotel, while the French, Italian, Spanish, Japanese and German suites will be redesigned in keeping with their original concepts.

A Presidential Suite will occupy the 10th-floor penthouse where the hotel's flamboyant creator and former owner, Victor Sassoon, once lived.

HBA's design for the Fairmont Peace Hotel will recall Shanghai's renowned art-deco heritage, combined with streamlined furnishings and state-of-the-art in-room facilities. "It will be signature HBA – a look that is luxurious, contemporary and endearing – but muted and tailored for this particular property," stated HBA Principal Connie Puar.

The ground floor, which was originally designed as a luxury shopping arcade, will be returned to its classic crucifix floor-plan, with revolving entrance doors on all four sides of the hotel.

The splendid octagonal glass skylight and an entire mezzanine level – which had been covered up for decades with gypsum board – will be revealed once more. Custom-patterned stone mosaic floors will echo the hotel's original deco-style tiling.

A soft "buff and blue-grey" colour scheme will enhance the hotel's original intricate cornices and coffers. Refurbished copper balustrades and light fixtures will be complemented by antique bronze and polished nickel. Authentic period styling will also include grey-vein marble accented with rich Noir St Laurent dark marble borders and walnut burl grain wood paneling that were popular during Shanghai's 1930s art deco heyday.

华尔道夫酒店

项目地点：上海外滩
设计单位：HBA

　　建于整整一个世纪之前十里洋场的上海总会大楼，经国际著名室内设计公司Hirsch Bedner Associates(HBA)精心修缮后重现昔日光辉，重开为豪华的上海外滩华尔道夫酒店(The Waldorf Astoria On The Bund)。

　　此地标大楼建于维多利亚时代，洋溢英式文艺复兴风格，曾是上海最显赫尊贵的社交场所，附设保龄球场、餐厅、生蚝吧、游戏室、发廊及两间酒窖，其长达35米的Long Bar，曾为远东最长的酒吧，一度成为城中佳话。

　　此幢历史建筑落于外滩2号，随着岁月的流逝而光华退去，多年来曾多次被改建，用作办公室、赌场、电影摄影场地，翻新前最后用途为快餐店。

　　上海总会大楼现已喜获新生，重现了昔日的神采与浪漫韵味；作为希尔顿国际酒店集团在亚洲隆重开设的首家Waldorf Astoria酒店，为豪华旅游创立了全新标杆。

　　HBA设计总监Ian Carr表示："这一项目对于HBA、Waldorf Astoria品牌以及上海市来说，其意义之深远，难以估量。""HBA能够委此重任翻修这一历史建筑，我们感到无尚的光荣，我们为此倾情投入。最大的挑战就是要在保留其历史风貌的同时，加入当代旅客需要的奢华便捷的设施与时尚元素。"

　　上海总会大楼堪称外滩建筑群里一颗璀璨的明珠。大楼的外观是典型的英国古典主义风格，白色的外墙非常醒目，三四层中间贯以华丽的爱奥尼克柱，南北两侧室壁凸出，五层上南北端有塔楼。新近开张的上海外滩华尔道夫酒店由两座大楼组成。原来的上海总会大楼现在被命名为Waldorf Astoria Club，这是一幢全套房的大楼，穿过一个庭院，可到达另外一幢现代塔楼。酒店共设271间套房和客房、三家餐厅、两个酒廊、一家酒吧、一家精品店、大堂吧、宽敞完善的宴会空间与设施、游泳池、健身中心及水疗房等。

　　HBA在大楼内布置有维多利亚时代流行的家具与装饰，所用布料皆为宝石色调，并缀以大量贵气的吊穗与镶边。HBA采用了较具现代风格的艺术珍品及布艺装饰，令室内空间隐约流露时尚感。一盏气势恢宏、维多利亚时期风格的水晶吊灯，从中庭古老的天窗上倾泻而下。而历经岁月磨砺的古董地板也得以保存，使得空间更具凝重的历史感。

Waldorf Astoria Club 生动重现了当年上海总会的著名酒吧 Long Bar。昔日，宾客于吧枱前所站的位置，能反映出不同的社会地位。可惜曾贵为闻名中外社交热点的 Long Bar，早已失落于日本侵华时期。HBA 凭借库存照片对其原有特色的清楚反映，恢复了从深色实木嵌板、白色云石吧枱，以至线条硬朗的深色家具等各种细部，从而把昔日的 Long Bar 再现在世人面前。

这幢历史建筑内，最为浪漫的空间莫过于宴会厅 The Shanghai Club Ballroom。这里经过巧夺天工的修饰后，不但保留原有的温暖实木护墙、暗黄灰泥墙身、精致天花板与墙边装饰，更沐浴在古式水晶吊灯散射而来的柔和光线中，展现出昔日难以想象的精致与细腻。此外，一块巨大的充满东方色彩的地毯覆盖整个用餐区，让人印象难忘。

大楼内 20 间套房则可谓整项翻新工程中最瞩目的亮点所在，Waldorf Astoria Club 的套房被酒店形容为"超越家居"的住所，实在当之无愧。

当年的上海总会已经喜获重生，华丽转身为 Waldorf Astoria Club，使外滩 2 号再度成为象征时尚格调、高雅气质的地标，就像她的前身一样，这里是一个汇聚优雅与品味，提供极致奢华享受的尊贵府邸。

Waldorf Astoria Hotel

A legendary Shanghai pleasure palace dating back to the city's 'Pearl of the Orient' heyday has been returned to its former glory by HBA/Hirsch Bedner Associates and re-opened as the Waldorf Astoria On The Bund.

The iconic English Renaissance-style landmark dating back to the Victorian era and was once Shanghai's most exclusive and prestigious social club with bowling alleys, restaurants, an oyster bar, games rooms, barber's shop and two wine cellars. Its fabled Long Bar, extending 35 metres, was once the longest bar in the world.

The iconic heritage building at Number 2 on The Bund had long since fallen from grace – over the years accommodating office space, a casino, motion picture sound stage and most recently a fast food restaurant.

But it is now reborn, recapturing the spirit and romance of its bygone era while creating a new benchmark in luxury travel for Hilton Hotel Corporation's first Waldorf Astoria in Asia.

"Restoring this forlorn property was a tremendous honour and true labor of love, the ultimate challenge of taking historical context and adding up-to-date convenience and style" said HBA Principal, Ian Carr.

One of the finest architectural gems along Shanghai's waterfront promenade, the whitewashed fa?ade is adorned with Palladium columns, intricate gables and a pair of sculpted rooftop cupolas.

Combining the all-suite building, renamed the Waldorf Astoria Club, with a modern tower connected via a courtyard, the prestigious new hotel features a total of 266 rooms, four restaurants, two bars, a patisserie, lobby lounge, extensive banqueting facilities, swimming pool, health club and spa.

Throughout the painstakingly careful restoration, HBA designers were mindful of maintaining the highest degrees of period accuracy, requested by the Shanghai Cultural Relics Management Commission which oversees the city's heritage buildings.

"That the Shanghai Club has its roots in the late Victorian era was never far from mind," said Mr Carr, who was also co-lead designer. "Restoring or converting heritage property is the ultimate design challenge."

Because of the Shanghai Club's wide popularity, the photographic archive was fortunately immense, guiding restoration of nearly every element from its neo-classical interior and English colo-

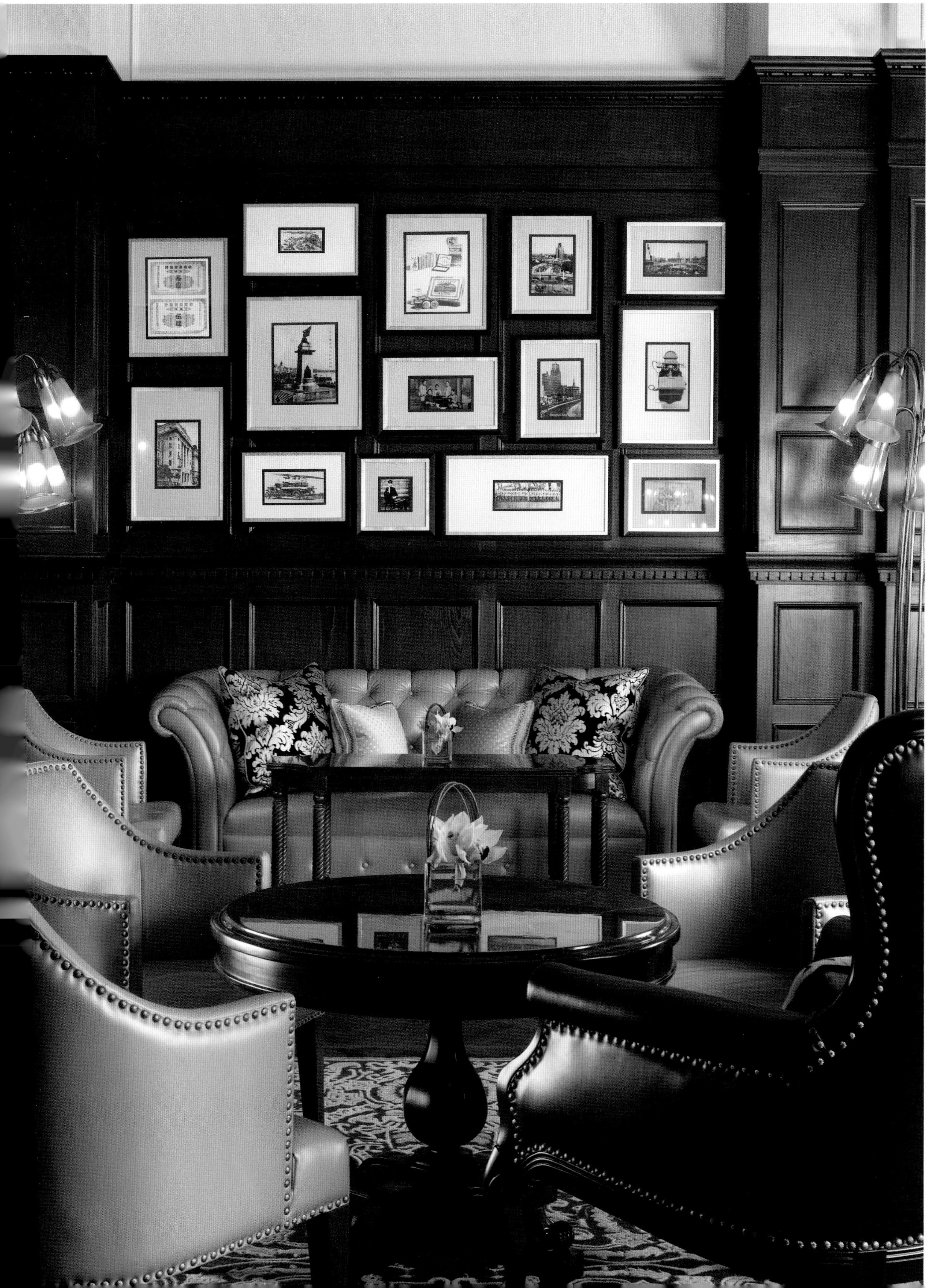

nial-inspired furniture to decorative lighting and antique Chinese accents and artifacts.

Interiors including original Sicilian marble columns and stained glass imported from Birmingham, England were painstakingly restored through extensive use of archived photographs and records.

Connie Puar, HBA Principal and co-lead designer, said: "Restoration demands extreme attention to detail combined with flexibility. When we find a column where we did not expect it, we have to amend in a way that won't compromise the design, which can be a challenge. Every item physically attached to the architecture—from ceiling molding and panels in walls, to the Sicilian marble columns and stained glass imported from England—is a restored or recreated heritage element."

After removing decades of accrued material, HBA found the original interiors to be quite enduring. Paula O'Callaghan, Associate and senior FF&E designer, said: "A good amount of original surfaces like false moldings and drop ceilings were there, hidden underneath movie props.

"There were a number of surprising superficial changes, including a scene of the Last Supper on the ballroom ceiling and finding the Shanghai Club's fabled Long Bar painted in KFC's palate."

A New Era

The Waldorf Astoria Club also resurrects the Long Bar. At one time, patrons could gauge their relative social status by where they stood along it. A world-famous destination, the original Long Bar was lost during the Japanese occupation, but is now recreated from photographic records to original specifications, from dark timber paneling and white marble counter to rich, dark masculine furniture.

The most intentionally romantic space in the heritage building, The Grand Ballroom has achieved a level of refinement simply unimaginable in its early days, bathed in light by heritage-inspired crystal chandeliers while retaining its original warm, dark wood wainscoting, buff painted plaster walls and intricate ceiling and wall moldings. A remarkably scaled Oriental style rug covers the entire dining area.

Twenty heritage building suites are perhaps the most impressive element of the transforma-

028

tion. The Waldorf Astoria refers to the Club suites as offering "better than home" for good reason.

HBA's design team were called on to imagine a room that might have satisfied visiting royalty in 1911, and succeeded in making classical European style more romantic and refined than any original could have been, with oversized marble bathrooms, walk-in closets and dramatic fireplaces.

Décor is sumptuous with exquisite chandeliers, dark mahogany period furniture and Chinoise, antique reproductions.

"Ultimate design of the Waldorf Astoria Club delivers an unhurried feel of refined indulgence; as timeless today as it would have been a century ago, and able to stand next to the Waldorf Astoria on Park Avenue on equal footing," added Mr Carr.

"In its rebirth as the Waldorf Astoria Club, Number 2 The Bund is again an address associated with high style, gracious civility like its cherished predecessor – a bastion of civility and style with the finest luxury experiences in Shanghai."

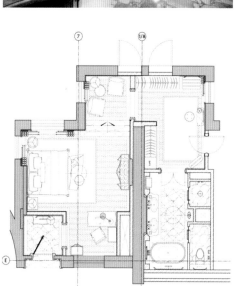
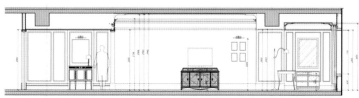
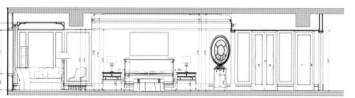

上海玻璃博物馆

在中国打造一座玻璃博物馆是一件兼具挑战且充满乐趣的工作。我们不仅可以每天都围绕着独一无二的玻璃艺术品工作,同时还能沉浸于悠久的历史文化,而玻璃又是如此多变,且有着多元的表现方式。从建筑玻璃到精密的科学仪器、到古文物、到当代设计作品,在工作中,博物馆团队启发了我们对拥有五千年历史的玻璃美感及技术的认识,一步步引导我们走入玻璃的世界。作为博物馆的设计团队,我们将自己视为第一批参观者。我们希望让每个参观者都了解展览内容,也就是用有趣且易懂的方式将信息转换成知识。最后所呈现的结果就是以多元的体验式环境结合互动的方式,让参观者亲自参与学习,逐步走入历史,了解古代及现代玻璃。

博物馆主展厅的参观以一个万花筒入口作为开端。万花筒入口的剖光不锈钢折面反射出5块大型视频上的影片,影片从抽象的艺术角度展现了"玻璃与人类生活"的主题。影片是由德国籍导演Nina Rose为博物馆量身定制,触及的题材包括老化、脆弱、欢乐和美丽等。紧随其后,是一张放大了的来自美国康宁玻璃博物馆的古埃及玻璃斯芬克斯的图片,以及源于西周时期的中国古玻璃珠。在这一部分的参观过程中,我们讲述了中西方的玻璃技术发展史,从两河流域直至上海的高新技术企业。

另一幅由瑞士著名插画工作室Jacques et Brigitte创作的巨型壁画,描绘了日常生活场景,并引发了诸如"玻璃能有多快?""玻璃能有多热?"的疑问。所有问题都能从隐藏在门后的互动式展品和多媒体装置中寻找到答案。从航天飞机的尖鼻处的高科技玻璃到一个普通的真空保温瓶,从生物玻璃到道路铺设使用的回收玻璃,玻璃存在于我们生活的方方面面。玻璃屋着眼于视觉设计,展出了一些当今国际知名的杰出设计作品,如Jamie Hayon为Baccarat所做的限量版作品以及VENINI为其90周年纪念所创作的"VENINI之柱"。玻璃屋的空间绝大部分将被特制的1000-2000个激光

项目地点:上海
建筑设计:德国罗昂建筑设计咨询有限公司
室内设计:协调亚洲
设计师:Tilman Thurmer
摄影师:diephotodesigner.de

刻字的银色玻璃瓶所占据。这些玻璃瓶将被安放在四周的玻璃展架上,而每个玻璃瓶上所雕刻的中文汉字也将组成一个美丽的爱情故事《玻璃山》。

博物馆二楼将作为现当代玻璃艺术展区,特别展示国际及中国知名艺术家,例如Steven Weinberg, Dale Chihuly 的作品。而在对面的回廊上将会安装一个长达25米的霓虹灯作品,内容是来自于Anton Chekov 一句著名而深刻的名言:"Don't tell be the moon is shining; show me the glint of light on broken glass."?

耐心,激情和执行力——经年累月的积累才能创建起一个优秀的博物馆和一系列非凡的藏品,甚至需要更多年的学习、观察以及聆听来对此做出评估,进行自我提升。上海玻璃博物馆不会只是另一家摆放几件展品做单纯展示的博物馆,上海玻璃博物馆会通过提供以教育和娱乐为导向,且具备当代国际水准的展览环境,成为吸引对设计和文化感兴趣的国际游客的新地标。

Shanghai Museum of Glass

It is a challenge and interesting work to create such a museum in China. We can enjoy the unique and special glass arts every day and immerse into the long history culture. In addition, the glass is changeable and has many diverse manifestations. From the architectural glass to the sophisticated scientific instruments, ancient artifacts, and to modern design works, during the work, the museum team has inspired our soul on the glass beauty and its technology which possesses five thousand years history. They lead us into the glass world step by step. As the museum design team, we consider ourselves as the first visitors. We hope each visitor can understand the exhibition, that is to use an interesting and understandable way to convert the information into knowledge. Finally, the exhibition will show a diverse experiential environment in an interactive manner so that the visitor can learn the history and know more about the ancient and modern glass.

The main exhibition hall of museum is an entrance of the colorful inner arts. The kaleidoscope entrance's stainless steel has reflected five huge block video movie. The movie has demonstrated

the topic of "Glass and human life" from art perspective. The movie is made by German Director Nina Rose for museum specially. It refers to aging, vulnerability, joy and beauty. Next to it is a Ancient Egypt Sifancours Glass picture coming form America Kangning Glass Museum, as well as Chinese ancient glass balls of Xizhou Dynasty. In this visiting process, we have told the glass development history of both China and western countries, from Two-rivers area to Shanghai High-tech enterprises.

The other huge wall picture is created by Jacques et Brigitte who is an famous drawing designer of Switzerland. This picture has shown a daily life scenery and also let people thought that "how fast can glass be?", "what a high temperature can glass be?". All these questions can be addressed through the interactive exhibitions and media device behind the door. From the high technology glass on airplane to an common heat-keep glass, from the bio-glass to the recycled glass used for paving road, glass exists in out life in different aspects. This glass room focuses on the visual design, showing some international famous design works, for example, the several works made by Jamie Hayon for Baccarat as well as "Pillar of VENINI" created by VENINI for their 90 years anniversary. The most space of glass room is covered by 1000-2000 laser word carving silver glass bottles. These glass bottles will be placed on the glass exhibition frame around. However, the carving words on each glass bottle will form a nice love story of "Glass Mountain".

The second floor of the glass museum will be used as the modern glass art exhibition area, especially for the well known artists from both home and aboard, such as works of Steven Weinberg, Dale Chihuly. However, at the corridor, there will be installed a neon lamp works with 25 meters long. The content comes from an well know saying of Anton Chekov.

Patient, passion and executive ability—after years accumulation, such an excellent museum and series of amazing works can be created. It is even necessary to learn, observe and listen to the evaluation and to improve ourself. Shanghai Glass Museum will not only show several exhibitions but to provide education and creation to people. It i possesses modern international standards of environment, which will become an new landmark for international visitors who are interested in design and culture.

上海外滩英迪格酒店

项目地点：上海外滩
设计单位：**HBA**

由享誉国际的室内设计公司 Hirsch Bedner Associates（HBA）一手打造的上海外滩英迪格酒店最近惊艳亮相上海。该酒店是洲际酒店集团旗下亚洲首家英迪格酒店，HBA 的创新设计兼收并蓄而又亲切和谐，体现了上海东西交融、海纳百川、面向未来的城市精神。

英迪格品牌的理念是呈献融汇本土特色的精品酒店，让宾客产生与当地社区紧密相连的亲切感。HBA 为了使这家英迪格亚洲旗舰酒店实现这一愿景，精心打造出了一家"拥有独特个性"的酒店，设计遍及酒店的 180 间客房，当中包括 21 间江景套房及两间宽敞的花园露台套房。

酒店的大堂入口异常绚烂瑰丽，堪称沪上一绝；既反映了酒店位于黄浦江畔的位置，还体现了品牌对自然环境、循环再用，以及生态敏感型设计的承诺。HBA 选择原钢、混凝土、外露砖及抛光石膏等富有张力的基本材料为大堂进行装饰，令人不禁联想到这一空间是从码头旁的滨江阁楼改建而来。而开放式隔室与清水混凝土天花便进一步增强这种效果，并配以全日色彩幻变的灯光。

与大堂如出一辙，客房也呈现一种自然色调：外露的上海灰砖、磨耗效果的灰色嵌板、抛光石膏墙和帆布。与之产生强烈对比效果的是色彩鲜艳跳跃的地毯。

中式灯笼、传统家具、陶瓷和古董等兼收并蓄、机巧别致的工艺品和家具带来老上海的感觉。带顶篷的睡床为原创设计，灵感源自传统中式婚礼所用的喜床，经当代手法重新演绎。

偌大的浴室设有一堵镶在抛光钢框中的玻璃墙，望向黄浦江；并设开放式湿区，当中附

设配上长方形瓷面盆的简约盥洗台，营造当代风尚；而独立浴缸也同样时尚摩登。

HBA这一创新设计古今交织，堪称奇迹。呈现在世人面前的是一个充满年轻活力、符合当代潮流、蕴含无限灵感的极致空间。这一空间从上海的历史走来，并将开创上海未来设计的新风尚。

HBA设计的这一亚洲首家英迪格酒店，使英迪格品牌雄踞上海外滩十六铺这一充满近现代历史风云的时尚新地标，并且为未来的英迪格酒店树立了一个可以借鉴的标杆。

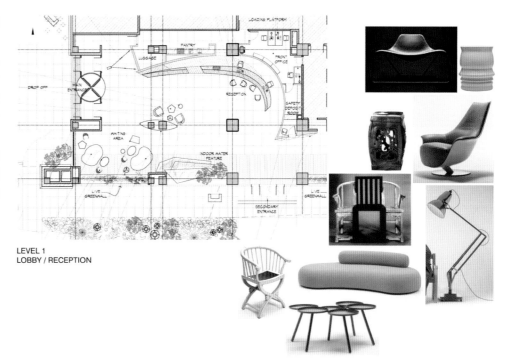

LEVEL 1
LOBBY / RECEPTION

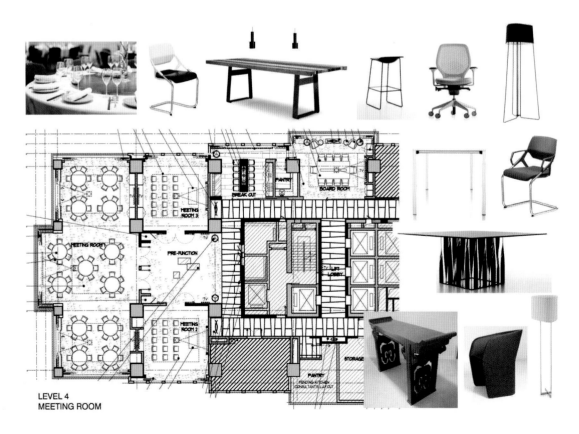

LEVEL 4
MEETING ROOM

Hotel Indigo Shanghai on the Bund

With the goal of transforming Hotel Indigo Shanghai on the Bund, flagship for the debut of InterContinental Hotel Group's boutique Hotel Indigo brand in Asia, global interior design firm Hirsch Bedner Associates (HBA) has stamped an innovative design onto the hotel that is at once eclectic and harmonious, a design that connects the ancient with the modern.

The philosophy behind the Hotel Indigo brand is to offer boutique-style hotels infused with local inspiration so guests feel connected to the local neighbourhood and community. To fulfil this vision at the Asian flagship, HBA's concept was to create a "personality all its own" for the 180-room hotel, including 21 River View Suites and two spacious Garden Terrace Suites.

"Throughout, Hotel Indigo's design is about connecting the hotel to the neighbourhood—one anchored by the river and its influence on commerce and connection," said Andrew Moore, HBA's lead designer on the project.

HBA developed an eclectic and harmonious design linked to the neighbouring Huangpu River, and the element that ties it to the neighborhood most intimately, Shiliupu Dock, now know as Pier 16. This dock was the gateway through which Shanghai grew, as a shipping and trade centre, and entry point for thousands of European expatriates

who led Shanghai's development as a global city.

"While a general sense of place is often a hallmark of thoughtful hospitality design, HBA dials in place at a whole new level of detail for the Hotel Indigo: not a nation or even a city, but a neighbourhood," added Mr Moore.

The lobby entrance is among the most striking and dramatic in Shanghai, reflecting Hotel Indigo's position on the river and the brand's commitment to nature, recyclables and ecologically sensitive design.

HBA chose strong elemental materials to render the lobby: raw steel, concrete, exposed brick, and polished plaster – suggesting this gallery space has been repurposed from a wharf-side waterfront loft. The open cell, cast concrete ceiling enhances this effect, studded with lighting that changes colours throughout the day.

As in the lobby, the guestroom palette is the natural tone of exposed Shanghai gray brick, distressed gray paneling, and polished plaster walls, a canvas against which shines colorful and lively carpets.

The sense of an older Shanghai is in eclectic and whimsical artifacts and furniture: with Chinese lanterns, authentic furniture, ceramic pieces and antiques. The canopy bed, an original design, was inspired by traditional Chinese wedding beds, but reinterpreted though a contemporary lens.

"We found wonderful furniture— for example, an especially interesting armoire console—in the local bazaars," said Mr Moore. "We had it re-

stored, then sprayed in fresh white enamel, so the piece would be simultaneously old and new." It became the model for reproductions used in each room. Other furnishings reflect ecological sensitivity: while the pieces vary room to room, the materials are all eco-friendly.

Oversized bathrooms have a glass wall framed in polished steel, looking out onto the river. They feature an open wet area, where a minimalist vanity topped with rectangular porcelain basins gives a contemporary feel, as does the freestanding tub, which is a sleek and modern.

HBA's innovative design accomplished a true feat: connecting the ancient with the modern. The result is a youthful, contemporary, inspired space that understands where it has come from and leads the way into Shanghai's design future.

In helping develop the first Hotel Indigo in Asia, HBA created a design that transforms the boutique brand name into a vantage point onto the most interesting areas within the storied and dynamic city of Shanghai – creating a standard against which all future Hotel Indigo properties will be judged.

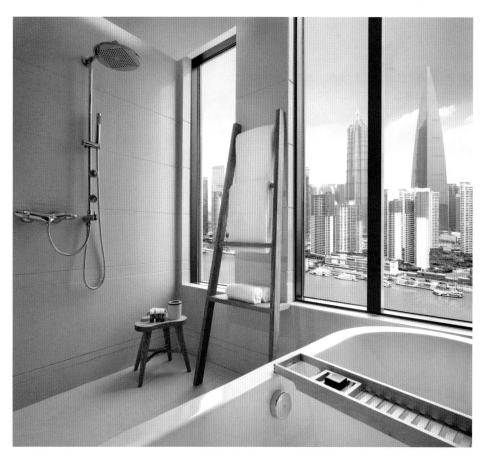

and··· - SuperPress - SuperBla办公室

项目地点：上海 八号桥一期
设计单位：Naco Architactures
设计师：Marcelo Joulia
建筑面积：80 m²
摄影师：徐文磊
撰文：Amber Chan

"Galaxy",对应的中文意象是:银河,星系,以及——一群杰出的人。初识and…,SuperPress, Superbla的人可能不能立刻了解这几家公司之间的关系,其实只要记得它们是一个"Galaxy"就好。

整个办公空间由纳索(Naco)建筑设计工作室设计完成。"白色"以"统领者"姿态霸道占领这个空间,让人想到北欧的皑皑白雪,茫茫无际状。所有颜色光线聚集在一起,最后集大成的便是——白,最单纯不过也最博大深广的颜色。纳索上海设计总监Margaux Lhermitte女士告诉我们纯白创造出解除禁锢感的空间,原初纯朴未经加工,像是一块为不设限的创意准备的空白画布,用"a-n-d…"的CEO——Ruth Ang的话来说,这里就像是"裁缝的剪裁工作室",激发灵感任人挥洒。

点缀其间的红色尤为醒目。红色圆形大地毯铺成第一层办公空间的圆心,环绕其周的鲜红椅垫,还有攀着旋转楼梯蜿蜒上楼的红地毯,像是贯穿整个空间的血脉。红色是拥有磅礴力量的色彩,热情热忱是它的专属语言,令人充满活力,将沉闷呆滞一扫而净。

"红白相间"是Naco和这个galaxy钟爱的色彩搭配,SuperPress的红白logo就是其中例证之一。

在色彩上还占有一席之地的是绿和紫。这两色的选择也不是临时起意,其灵感来自以极简美学著称的时尚设计师Jil Sander的时装发布会,它们也都拥有旺盛的能量,浓艳鲜妍。很难在这里的"绿"和"紫"前面加上描述的定语,很难用"墨绿"、"翠绿"、"草绿"、"青绿"或是"茎紫"、"晶紫"、"红紫"、"浅紫"来定义它们,因为它们的颜色层次随着时光流转在人们不经意间悄悄变幻。"因为创意一直都在变。"Margaux说,"那颜色为什么不?"

最让格子间总是抱怨苦闷的白领们眼前一亮的,或许是这里造型各异独特新奇的办公桌。虽说造型各异,但也并非无迹可循,它们像是原本完整的一大块实木,用半径不同的圆模挖空,也像是裁缝剪过圆形布块后剩下的边角料。这些"边角料状"的办公桌在Naco的鬼斧神工下,华丽转身为一个特别的存在——"让在里面工作的人既可以轻松自在相互谈话",随时随项目变化团队方阵,又能让每个人拥有自己独享的专属空间。打破约束和隔阂,是Naco为这个空间设计自始至终的坚持点。就是"有机感(Organic feel)",这是Ruth用的词。

旋转楼梯是Ruth的心水所在。在设计上,这个旋转楼梯贯通上下空间,打破平面的单调,创造出另一个新的工作氛围。它不仅在外形上以灵巧蜿蜒大得分,最特别的地方是——楼梯下通常是废弃的空间,本案中被镂空为一个"我们的骄傲"展示台,不用来放包放杂物,有资格登上这些展示台的都是值得让他们感到骄傲的作品。

"审美意识"在这里有着非凡地位。小到台灯,纸篓,大到挂壁艺术品,都经过精心挑选,因为它们都承载"激发灵感"的重任。不得不提的是旋转楼梯上空悬的白色大鸟笼,常令访客感到惊奇。"这正好就是我们希望达到的效果:惊讶——惊奇——惊喜。"鸟笼为这里营造出更丰富的情绪,类似于孩童时代的嬉戏情怀。

但是,在这一个完全开放式的空间,私密性要如何解决?

其实这也是有意为之。Naco和在这个空间工作的团队都不希望设置任何封闭或隐蔽空间,因为"头脑风暴"是他们的每日必修课,他们本来就应该持续在团队中工作,相互间以及和客户间交换各自的想法和创意。

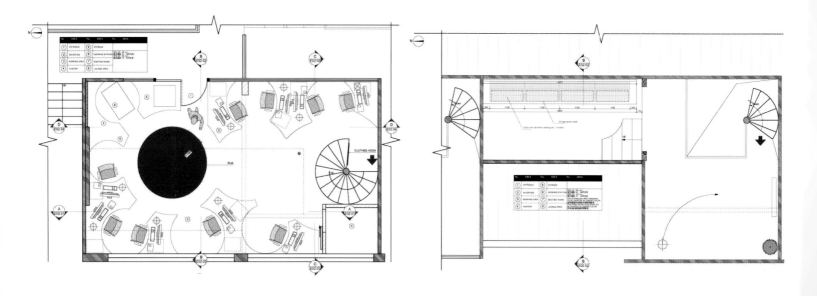

SuperPress SuperBla Office

"Galaxy" means Milky River and galaxies, as well as —outstanding people. The people who acquaintance with SuperPress, Superbla may not understand the relationship between them. You only need to remember "Galaxy".

This office is designed and completed by Naso Construction Design Studio. "The white color" is covered this space in a "dominate" way, letting people imagine of the endless snow in Nordic. All the colors and lights focusing here have formed a single performance—white, which is the most purity and greatness color. Ms. Margaux Lhermitte, the Design Director of Naso in Shanghai, told us that the white has created the most freedom space. This pure color looks like a bland canvas with endless creative. In CEO—Ruth Ang's word, it is just as a "tailor studio" to inspire people's soul.

The red color is dotted in the space. The round and red carpet has paved the first floor's space, surrounding with the bright red cushions and the red carpet on the spiring stairs, which looks like the veins of the entire space. Red is a majestic force color with enthusiasm. Enthusiasm is its exclusive language, letting people fill with vitality and let the boring go.

"The red and white color" is the favorite color adopted by Naso and galaxy. The red and white logo of SuperPress is one example.

The other important colors are green and purple. These two colors are selected not by moment. Its inspiration comes from the fashion designer Jil Sander's Fashion Collection. They are possessing the genetic power, brightly and colorful. It is hard to add description words before "green" and "purple". It is hard to add "ink green", "bright green", "grass green", "light green" or "dark purple", "crystal purple", "red purple" or "light purple" to define them, because their color are changing slightly with times. "Because the creation is always changing", Margaux said, then, why not color?

"The Best Grid" is always surprised by the boring whiter colors, maybe because there are different shapes and unique strange office table. Thought they are different shapes, they have certain rules. They are looking like an original complete wood and be hollowed with different diameters. They are also looking like the rest parts of a piece of cloth cut by tailor. These "cast-off" shape office tables are special and unique through great skills of Naso—"letting the workers can have a easily communication". These tables can be changed into different style in accordance with the different projects and let each people possess their own special space at the same time. Breaking through the constrains and separation is a great point insisted by Naso for this space. That is "Organic feel", which is described by Ruth.

The spiring stairs is the focus of Ruth. On the design, this spiring stairs passes through the whole space between ups and downs, which has broke the boring plain and created a new working atmosphere. It is not only unique in terms of the flexible out shape, the most great point is that the

under part of stairs is always abandoned and the designer hollowed it as a "Proud Exhibition" platform. It is not used to store the bags and packages. The products that can be displayed on are all products we proud of.

"Aesthetic consciousness" has a major position here. From the lamp, wastebasket to the wall-hanging art, they are all selected carefully, because they are all shouldered the responsibility of "inspiration". The one I have to mention is the hanging white birdcage on the spiring stairs, which brings visitors surprise often. "It is just the effect what we want to express: surprise—amazing—comfortable." It doesn't need to shaped in accordance with the traditional standards. The birdcage here is to create more fun, which is looking like the childish sense.

However, in this completed open space, how to reserve a private space?

In fact, it is also set by purpose. Naso and this work team are all not hope to set any closed or secrecy space, because "brain storm" is the necessary class for them each day. They of course shall work for this team and exchange their own ideas and innovations with customers and themselves.

隐泉之语

项目地点：上海
设计单位：Imagine Native Ltd. (Hong Kong)
建筑面积：320 m²
摄影师：Kingkay Architectural Photography

本案位于上海国际金融中心,设计师利用折纸为整个设计概念的主要驱动力。餐馆空间划分为不同的子区域,寿司吧,酒吧,两个主要餐饮区,卡座和榻榻米房间。每个区域的空间是由一个像折纸的特色建构设计而成,而这建构是由不同颜色和材料组成,如多孔铝复合板,透光石面板和麻质布料。加上不同的灯光效果,每个区域都将有自己的空间特征。这些不同的子区域也由一折纸吊顶连接起来,从入口一直到在店尾的贵宾房,它整合这些功能子区域,并将整个餐厅的空间转变连系。

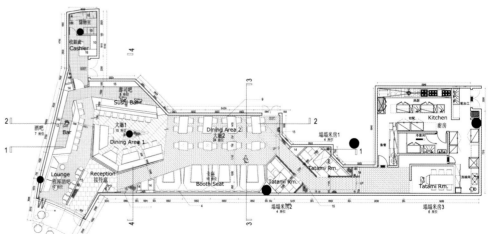

Haiku Sushi

The new Haiku Sushi is located at the open courtyard of the recently completed Shanghai International Finance Center. We utilize the concept of origami as the major driven force throughout the design. The restaurant space is sub divided in different zones, sushi bar, drinking bar, two main dining areas, booth seating and tatami rooms. Each zone is formed by an origami feature, which is constructed by colors and different materials, such as perforated aluminum composite panels, translucent stone panels and linen fabrics. With different lighting effects, each zone will have its own spatial character. These different zones in the restaurant are also tied by an origami suspended ceiling from the entrance to the VIP rooms at the end of the shop. It unifies these feature zones and creates a spatial transition throughout the restaurant.

Kartel 酒吧

项目地点：上海襄阳北路1号5楼
设计单位：Lime 388
设计师：Thomas Dariel、Benoit Arfeuillere
主要材料：定制瓷砖、地板、玻璃、油漆
建筑面积：四楼 183 m²；五楼 186 m² 天台 186 m²
摄影师：Derryck Menere

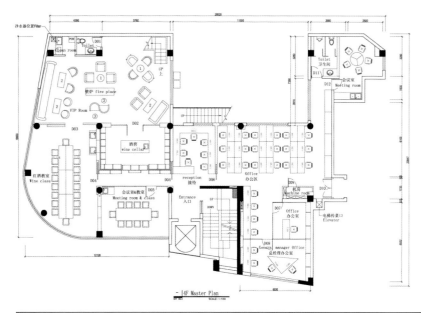

Kartel 是一座豪华酒吧，位于前上海法租界的核心地段。该地段是一个宜人的居住和购物区域，中欧风格的建筑，国际餐厅和小商铺，梧桐树成荫的街道和老式弄堂在这里完美融合，这就造就了今天的上海。随着"摧毁别致"的概念，两位设计师 Thomas Dariel & Benoit Arfeuillere 希望使 Kartel 在这片充满历史记忆的地域里，展现其优雅而又发人深省的强烈对比。

由于 Kartel 所在的大楼仍在进行之前的拆除工作，这更自然地契合了"摧毁"这一主题。伴随着破旧墙面的剥落，大楼本身的结构逐渐显现出来，更突现了应当被保留的形状、质地以及深厚的历史记忆。在欧洲，"Beaux-Arts"风格向来推崇那些剥去一切掩盖其原始建筑的楼宇，因此设计师决定在此也顺势保留几处原始的风味。剥落的墙壁，未包裹装饰的柱子，建筑工人留下的字迹，都成为这些原始风味的历史元素。为了平衡这一点，设计师特别运用优雅醒目的装饰和时尚的定制家具来营造 Kartel 充满欧亚风情的温暖氛围。酒吧一共有3层，其中两层位于前法租界的一幢老楼的4-5楼，外加一个拥有360°环景的露台，这三层不同风格的空间分别演绎了不同的情调氛围。

4楼的布置和壁炉令人想起传统巴黎沙龙的温馨气氛，喝上一杯勃艮第红酒，与三五知己时而热烈时而亲密的聊天。5楼是酒吧的主要

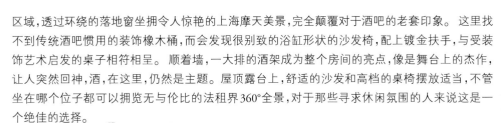

区域，透过环绕的落地窗坐拥令人惊艳的上海摩天美景，完全颠覆对于酒吧的老套印象。这里找不到传统酒吧惯用的装饰橡木桶，而会发现很别致的浴缸形状的沙发椅，配上镀金扶手，与受装饰艺术启发的桌子相符相呈。顺着墙，一大排的酒架成为整个房间的亮点，像是舞台上的杰作，让人突然回神，酒，在这里，仍然是主题。屋顶露台上，舒适的沙发和高档的桌椅摆放适当，不管坐在哪个位子都可以拥览无与伦比的法租界360°全景，对于那些寻求休闲氛围的人来说这是一个绝佳的选择。

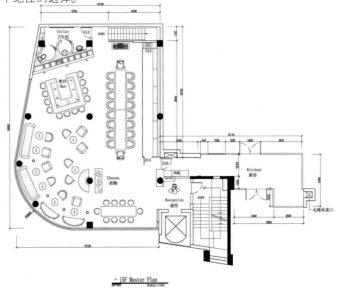

Kartel is a three-story lounge bar located in the former Shanghai French Concession. The neighborhood is a pleasant mix of residential and retail areas, Chinese and European style architecture, international restaurants and tiny stalls, wide tree lined streets and traditional lanes … a perfect combination of what constitutes the city today.

With a concept of "Destroy Chic", the two designers Thomas Dariel & Benoit Arfeuillere wanted with Kartel to play with these contrasts, to be elegant yet provoking in this very heritage district.

As the demolition of the previous space was going on, the "Destroy" part of the concept naturally grew on. While the walls were falling, the original structure of the building reveals itself, featuring shapes, textures, memories that ought to be kept. As the "Beaux-Arts" style in Europe pays homage to the buildings by ripping off everything that is hiding their original architecture, the creators decided here to keep several places as original. Exposed concrete walls, undressed stripped down pillars, and old letterings made by the people who built the place are part of these heritage elements.

Answering and balancing this raw side, Kartel offers an elegant and sophisticated décor served by comfy and stylish custom-made furniture, a warm atmosphere combining and mix-matching European and Asian influences. The lounge spans on three floors, two indoor plus a rooftop terrace. Three spaces proposing three distinct styles and atmospheres.

半岛1919红坊艺术设计中心

项目地点：上海宝山1919创意园
设计公司：吕永中设计咨询有限公司
设计师：吕永中
参与设计：席佳，区润宇，尹秀敏
建筑面积：1400 m²
项目日期：2011
摄影师：吴永长

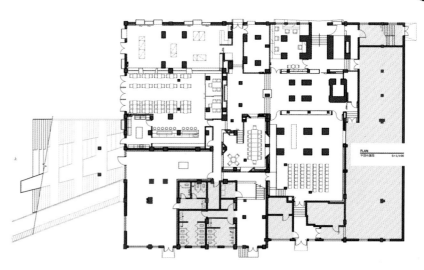

本案位于上海宝山1919创意园10号楼，占地面积约1400 m²，作为面向园区的复合功能服务平台，艺术中心拥有展览、会议、阅览、销售等功能组合，以满足园区内外不同需求。

园区前身是上海棉纺八厂，有近一个世纪的历史，10号楼则是前厂区的火力发电机房。先将煤运至楼层高处，再源源不断供应给底层的火力发电机组。根据运输、承重的实际需要，建筑底层密密麻麻建造了众多大小不一的巨型水泥柱，这是一个典型的构筑物空间。

一千多平方米面积有近百根柱基。经过分析，在原杂乱无章的梁柱中小心梳理，根据原结构使用功能，区分出建筑结构与设备基础。剥去设备基础外的粉刷涂层，呈现出水泥基础的原始粗犷力量，并小心保留工业历史时期的标识和痕迹，以最大程度还原其本质面貌。

粗壮的水泥立柱与狭小低矮的空间形成有如古堡式的封闭压力与探索的神秘。空间布局顺其自然，将大小不一的空间进行一系列精巧的串联，充分利用地面的高低起伏，隔墙虚实结合，灯光若隐若现，给人几分寻觅的期待。

把阅览室设置在中心区域，围绕四周布局其余功能空间。穿越四周低矮空间，登上有如圣殿般的"天井"阅览区。仰望着人造天光从9 m高的天井倾泻而下，并在墙面有序变化的冲孔木板上演绎出和谐、静逸的韵律。浅木色基调中晕染出温馨、淡雅的氛围，喻意着对知识和思考作为创意产业的活力与源泉的一种尊重和景仰。

带有方洞的轻薄隔断有如"纸窟"，轻柔附着于粗犷的水泥构件上，随着功能与空间转换延展，并提供了展示的功能需求，在历史的痕迹中翩翩起舞。

没有推腐拉朽式的拆除、没有光彩夺目的装点、没有华丽浓艳的涂抹，设计师对历史的欣然尊重，对空间细致入微的理解，谦和自然的创作态度以及简洁精炼的手法，让老的建筑得以延续着历史的积淀，并焕发出充满想象的新生。

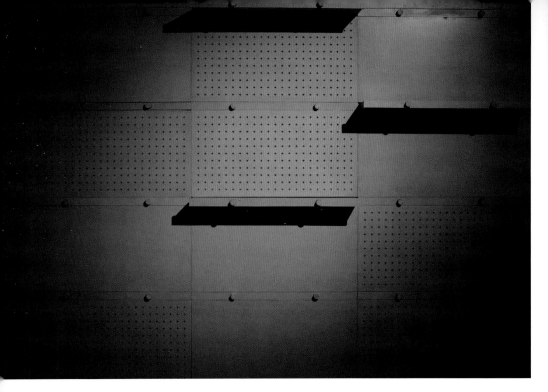

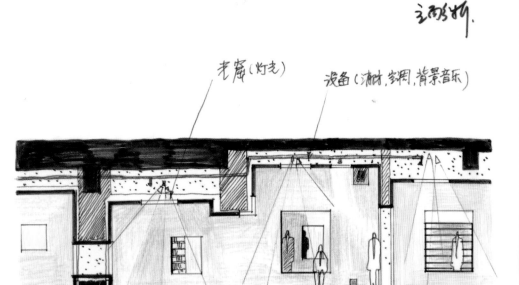

Designed and transformed by Bandao 1919 Red Workshop Art Design Center

This case is located at the No.10 building in Innovative Park of Baoshan Shanghai in 1919, covering 1400 square meters in terms of area. As the comprehensive service platform facing this park, this art center merges exhibition, conference, reading and selling into a single whole to meet the various needs in this park.

The former of this park is Shanghai Cotton Spinning No.8 Plant, which has a history with almost one century. The No.10 building is the thermal power general room of the first part of the plant. Firstly, they convey the coal to high level, and then put them to the thermal power generator continuously. In accordance with the actual needs of transportation and load-bearing, there are set various of huge cement pillar at the bottom of the building, which is a typical construction space.

There are almost a hundred of pillar among the one thousand square meters. Through analysis, among the disorderly pillars, according to the original structure function, they can be distinguished clearly in terms of construct structure and the equipment base. Stripping off the pink paint of the equipment base, it shows the roughness power

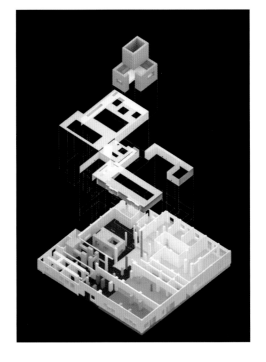

of cement. They have reserved the marks and tracks of industrial historical period carefully, which have re-shown the basic shape mostly.

The rough cement pillar and the low & narrow space have formed castle type's closing pressure and mystery. The space layout is in line with the nature, which has connected the different size spaces into together. Fully utilizing of high and lower space, partitions, dreamy light, they have brought some expectation to people.

Putting the reading room in the center part, the designer arranges the surrounding space. Through the surrounding low space, it gives you a palace-like "well" reading area. Looking up the man-made sunshine which pours down from 9 meters high, it performs the harmony and quiet rhythm combined with the changeable wooden plate on the wall. The lighten wooden tone has generated the warm and elegant atmosphere, which shows the respect and warship to the knowledge and consideration.

The light and thin separate wall with holes seems like "Paper Cave", which is attached on the rough cement softly. They change and extend with the function and space. In addition, they have realized the demonstration function, dancing in the historical traces.

No decay strike, no luxury decoration, no gorgeous painting, the designer adopts respecting to history, understanding the details of space, humble nature creating attitude as well as simply skills to let the old building succeed the historical traces and show the new appearance full of imagination.

域——中国室内设计年鉴

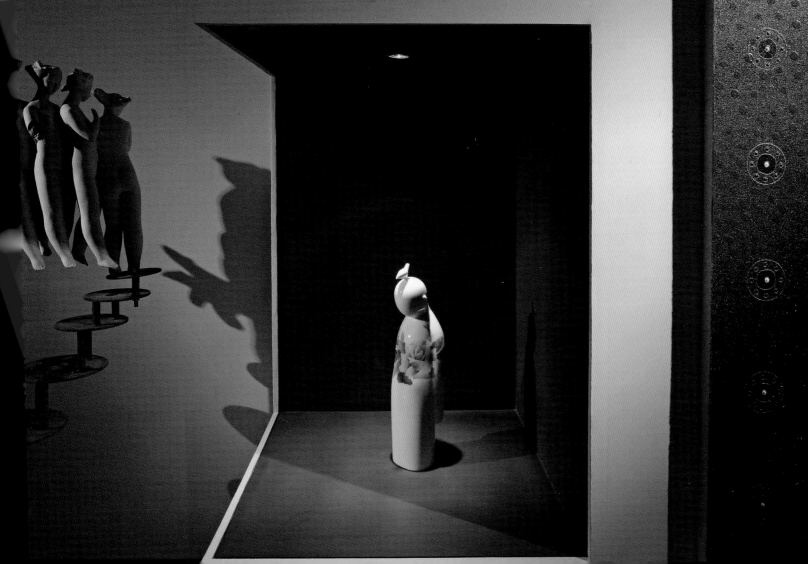

雷迪有限公司办公室

项目地点：上海天山路780号
设计单位：设计集人（www.designsystems.com.hk）
设计师：林伟明、梁芬华、杨励莹、张星、王永健、张芷茵、方欢欢、李婉恩
客户：雷迪有限公司
建筑面积：974 m²

这个办公室项目楼高5层，是一家地下管线管理及检测公司位于上海的办公大楼。

大楼拥有逾40年历史，属典型旧式现代主义建筑。由于当年采用的是讲求快捷及低成本的旧式建筑方法，即先预制好大部分水泥组件，再于现场装嵌，所以，整幢楼宇的所有墙身皆作负重之用，所有间隔皆不得改动，亦因此对这个项目的空间配置造成限制。经过力学计算之后，我们在负重要求最小的顶层拆除部分墙身，并加设钢架承托楼顶。所有楼层维持原来"中央一条通道，两旁布满小房间"这个井然有序的间隔。

一座富有历史的建筑物是记忆的载体。项目所在地原为牛类养殖场，后来成为农业原料国企之办公大楼，外墙与室内装修已沿用多年。其位处住宅区，毗邻上海特色的民房。为了使大楼与周边社区及环境互相协调，设计师选择保留大楼原有的窗户、楼梯扶手栏杆、墙身等这些饱经风霜的历史见证作为框架，以简单的白色油漆进行翻新，让大楼在隐然褪色的岁月痕迹当中，细诉昔日往事。

这个办公室项目是一家地下管线管理及检测公司，专门以先进的高科技仪器及方法如微型机械人等为水、电、热、通讯等各项公共事

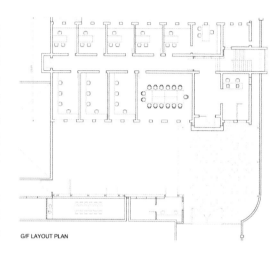

G/F LAYOUT PLAN

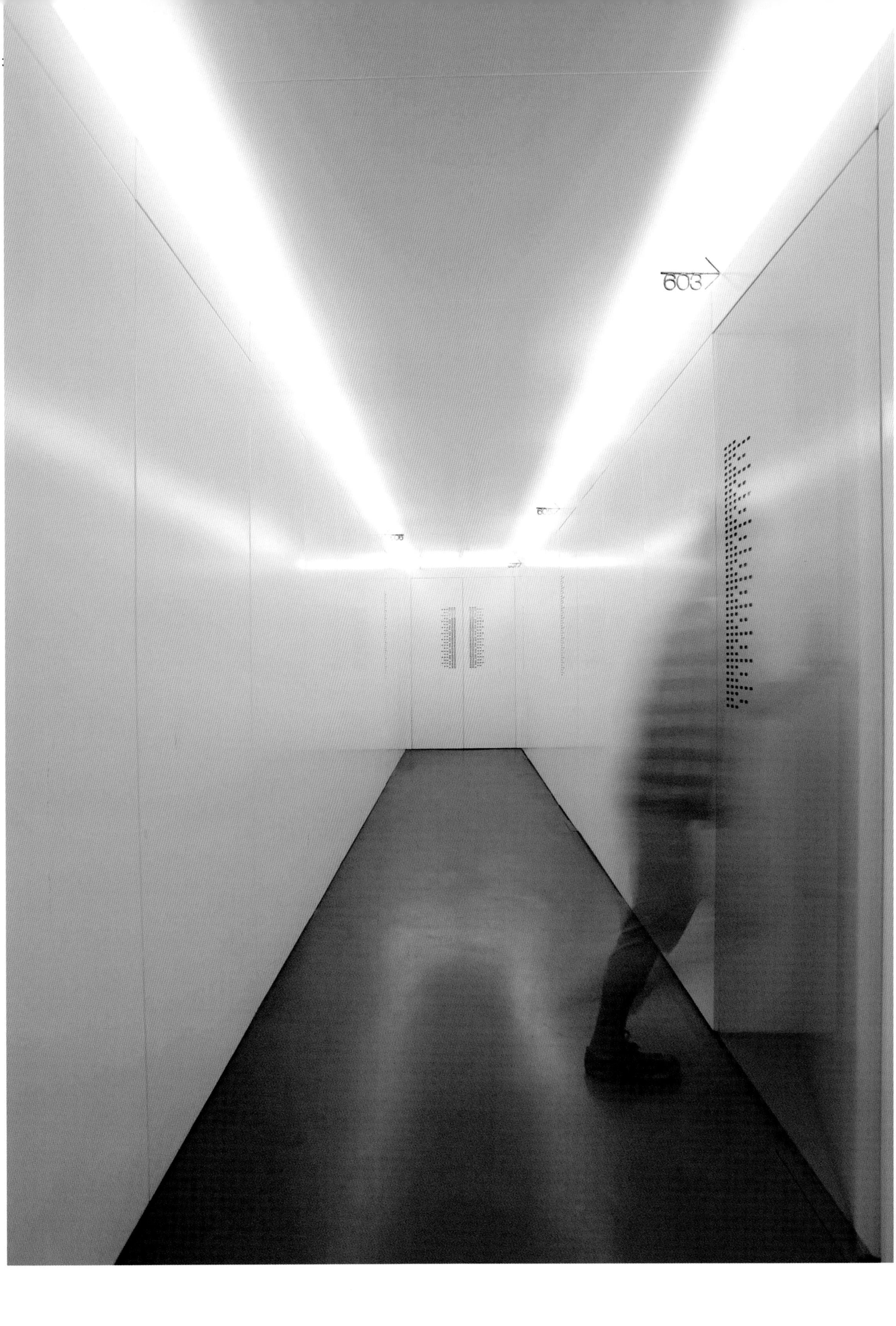
603

业提供地下检测服务,以便城市基建项目进行工程勘探和制定施工方案。由于这家公司的服务宗旨正是"看见你所看不到的",所以,设计师在这次的办公室大楼里特别设计了一条现代简约的主走道,各个不同功能的房间被巧妙地隐藏于两旁,同时房门又透过微妙的细节于走道间若隐若现。

这个项目的设计意念是怀旧老上海,设计师以铜作为设计语言贯穿大楼各个楼层。所有标识指示、门框及装饰皆以铜特制,令空间在焕然一新的同时,淡淡地流露老上海的独特风韵。窗外景色尽是充满老上海情怀的旧式海派民居;怀旧的设计正好将室内空间与窗外世界联系起来。

由于大楼内的楼层高度和空间间隔较小,各楼层办公室的天花均以铝材特别制成不同造形,以符合功能上的声学要求。如会议室的凹凸造形冲孔铝板天花,既具吸音效果,又能避免使用者直接看到光源,从而令室内光线更加舒适。

客户是一家中国公司,有不少外国的客户及供应商经常到访交流。为了强化这家公司的身份,董事长办公室内设置了一张云石茗茶桌,象征富贵的牡丹图案地毯,以及寓意平衡之道的铜制吊饰,来反映公司的文化背景与内涵。

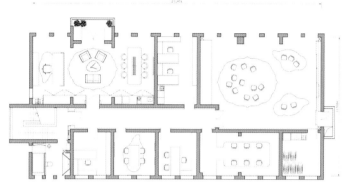

5/F EXECUTIVE FLOOR

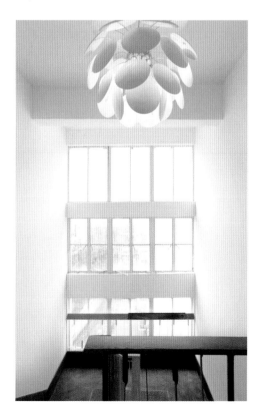

The office of Leidi Limited

This is a 5-storey office design project for an underground utility management and detection company in Shanghai.

This office block is a typical modernistic architecture of over 40 years old. It was built by assembling precast concrete slabs onsite, which was a popular building method at that time due to high efficiency and low cost. In this way, all the walls in the building are load-bearing, and no partitioning can be modified. This thus becomes the limitation on the space allocation of this project. After mechanical analysis, we removed some of the walls on the top floor and added a steel frame to support the roof. The tidy partition of having "one central corridor with small rooms on two sides" remains on all levels.

A building full of history is a vehicle of memory. The site was originally a cattle farm. It then became a national agricultural supplier's office building, of which the fa?ade and interior has been used for years. The building is surrounded by typical Shanghai houses in a residential area. In order for the building to be in harmony with the local community and surroundings, the designer has chosen to keep historical elements like the original windows, staircase handles and walls as the framework and repaint them white, so that the story of the building unfolds quietly through the pale traces of time.

The client is an underground utilities management and detection company. They specialize in providing underground detection services for public utilities and urban infrastructures by using high technologies like micro-robots. The company's philosophy is "seeing the invisible". Therefore in this project, a modern and minimal central corri-

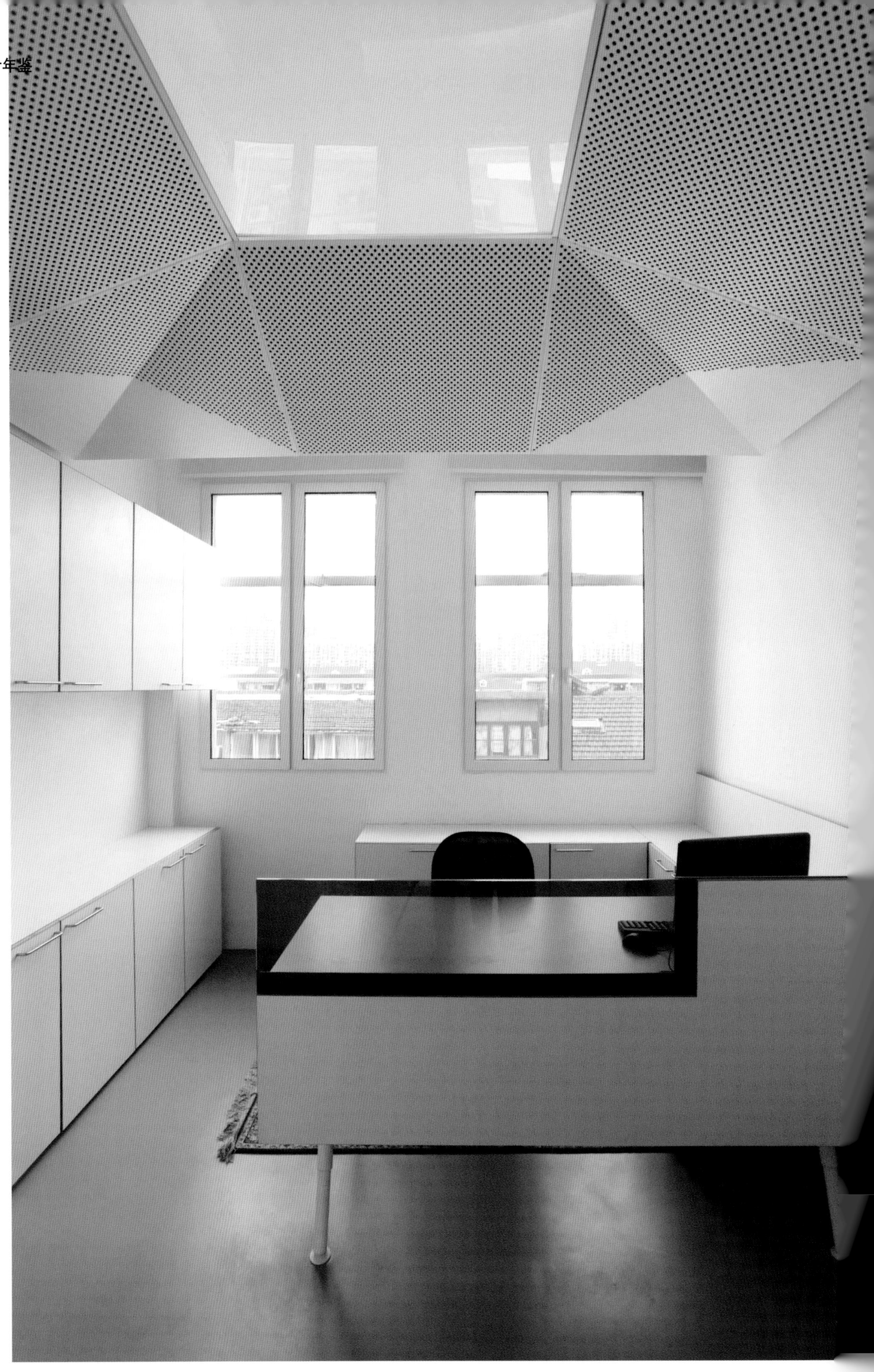

dor is specially designed in which the rooms of different functions on the two sides are neatly concealed, while the doors to these rooms are subtly revealed.

The idea of this project comes from Old Shanghai. The material brass is used as the thematic design language. All signage, door frames and decorations are made of brass to revitalize the space as well as to gently manifest the unique charm of Old Shanghai. Outside the window are typical old Shanghai-styled houses. The retro design exactly links up the worlds in- and outside the building.

Since the height of each floor and the space of each partition is relatively small, the ceilings of the office rooms are made of custom-designed aluminium profiles to functionally fulfil the acoustic requirement. For example, in the meeting room, the undulated ceiling made with perforated aluminium profiles not only absorbs sound waves, but also cleverly hides the light source from direct sight to produce comfortable illumination.

The client is a Chinese company with many foreign clienteles and suppliers visiting them. In order to strengthen such a unique identity of the company, details with Chinese characteristics are designed to reflect the company's cultural background and values, such as the marble tea table, the peony-shaped carpet that symbolizes "prosperity", and the copper hanging decoration that signifies "balance".

PARK 97 钢琴吧

项目位点：上海
设计单位：ANS International Design & Consulting Pty Ltd
设计师：徐岭啸、杨洁筠
建筑面积：348 m²

位在复兴公园的97俱乐部的Muse开业以来已经如此成功,因此业主决定相应扩大空间,同时增加2个私人贵宾室和1个大宴会厅称为"钢琴酒吧"。

顾名思义,一个白色的小型三角钢琴为主要特点置放在空间的中心。设计风格是侧重于纹理,光照和材质给它一个成熟而奢华的优雅气氛。整个空间通过走廊将其建立一个连贯的扩展。走廊的墙壁是附有质感的灰色块花岗岩和白色的数字,在黑色的背景,具有不同的音乐家和舞蹈家剪影。在钢琴酒吧,有黑色光泽亮面竹纹理墙壁环绕的房间。立体马赛克瓷砖,哑光灰色,与丝绸质感的黑色大理石,则是酒吧区和DJ区的主要特色。天花板是由不锈钢带相对应地板的木材和石头排列弯曲而成的形状。加上黑巧克力色,按钮簇绒沙发相结合,这个贵宾区展现了又一激动人心的经典却不失优雅的上海娱乐场合。

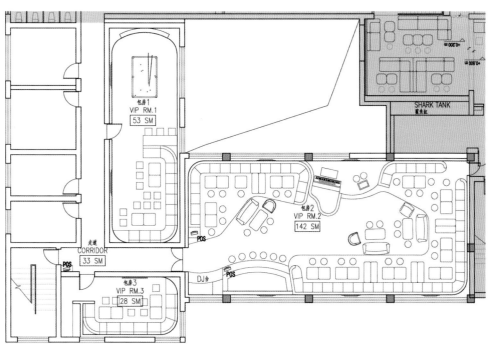

PARK 97 PIANO BAR

The Muse club at Park 97 club in Fu Xing Park has been so successful since its opening that the owner has decided to expand and add 2 private VIP rooms and 1 large party room known as the "Piano Bar." As the name suggests, a white baby grand piano is featured in the center of the space.

The design style is focused on the elegant play with texture, light and material to give it a mature yet lavish atmosphere. The corridor leading from the existing space to these rooms are also designed to create a coherent extension throughout. The walls of the corridor is textured with random grey granite pieces and white figures in the black background features various silhouettes of musicians and dancers.

In the Piano Bar, black glossy lacquer paint is coated on the bamboo-textured walls surrounding the room. Matte grey, three-dimensional mosaic tiles, combined with a silk-textured black marble are featured at the bar area and DJ booth area. The ceiling is decorated with chrome stainless steel strips cut with the curved shape that corresponds to the floor pattern of wood and stone.

Combined with the dark chocolate brown, button-tufted sofa, this VIP party area provides yet another elegant and classic spot for the exciting Shanghai entertainment and clubbing scene.

五维茶室

项目地点：上海市杨浦区军工路五维创意产业园区内
设计师：袁烽
参与设计：韩力、何福孜
业主：上海创盟国际建筑设计有限公司
建筑面积：300 m²
设计时间：2010.03-2010.08
建造时间：2010.08-2011.05
摄影师：沈忠海

位于创盟国际J-office办公区后院的茶室是对基地上原有的一栋屋顶已经塌掉的仓库房的再建。基地本身极为局促,三向面墙,另一个方向朝向一个有水池的后院,同时整个建筑对空间的索取也因为现有的一棵大树而受到很大限制,而设计的结果也表现为一种综合了封闭与开敞、占有与妥协、趣味空间与逻辑建造等多种复杂关系之后的一种和谐。整个建筑贴合基地空间,平面布局呈现为一个逻辑关系模糊的四边形,却也因此获得了对空间的最大索取。整个建筑在布局上分为三部分,朝向后院一侧布置相对公共性的开敞空间,一层茶室,二层图书室,同时在二层图书室伸出一个三角型的小平台将现存树木加以包裹,使得树木和建筑本身融为一体。而背向后院一侧布置休息室,书房以及辅助服务空间等相对私密的空间;公共空间与私密空间之间通过一个趣味性的连接空间得以串连。

连接空间是一个通过扭转放样得到的非线性六面体,将前后两侧的四边形平面去掉后,连接空间将两侧属性不同的功能空间加以融合,楼梯空间的置入同时解决了茶室的竖向交通问题,并为二层书房贡献了一个可以看到现存树木的内向小庭院。而连接空间也为本

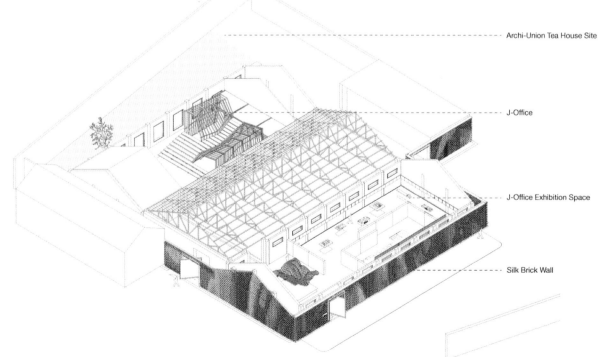

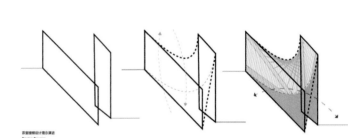
茶室楼梯设计理念演进
Design Process

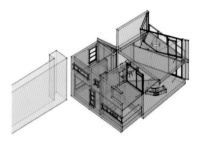

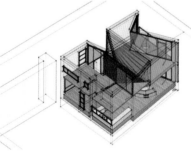

来平淡的基本功能空间创造了新的空间感受，一层茶室空间出现了从平直空间向竖向空间的突变；而二层图书室的空间也因为趣味空间的存在而获得了独有的场所感。

连接空间是一个无法通过平面图纸表述的三维异形体，我们在Rhino中完成的对形体的基本推敲以及空间的把握，但这样的数字模型很难直接转化为可以指导工人进行施工的讯息。同时工人手工施工的现有限制条件迫使设计师在提交施工方案时必须同时给出决措施，以实现前沿数字化设计与中国本土低技施工现实的结合。我们首先在数字软件中将曲面扫掠过的多根结构骨架线进行提取，使得曲面形式通过相互交错的直线进行概括，再将直线进行等分以实现直线间的曲面拟合，等分的距离控制在木模板可拟合的尺寸之内，这样数字化的放样就转化为手工可控制的形态。

再根据这样的直线拟合关系制作一比一的木骨架模具，在这一骨架基础上蒙上细分后的木模板，由此形成一个完整的空间曲面模板构架。模板构架根据施工工序切分为上下两次搭造。楼板的浇筑基本和普通的混凝土相一致，唯一就是钢筋的铺设也与模板的直线取向相一致。铺设钢筋后混凝土的浇筑也通过手工来完成，并最终形成了现有的实体效果。模板的痕迹在施工后完全保留，全手工的施工模式使得混凝土的表面出现了很多类似起泡、模板脱胶、钢丝外露等质量缺憾，但曲面的独特形式却使得这些得以弱化。虽然无论是模板布置还是手工浇筑都具有一定的造型误差，但这一数字化设计与低技手工施工相结合的方式对建造数字化建筑的探讨也具有了特别的意义。

Tea House

The Tea House located in the backyard of the Chuangmeng International J-office area is a re-shape building of a waste warehouse. The base itself is much limited. There are walls in terms of three directions, and only one is facing to a pond in the backyard. At the same time, the entire building is limited to the huge tree in terms of space. Therefore, the design is also showing a complex but harmony appearance that combines closed and open, covered and compromise, funny space and logical construction and so on. The entire construction plain is showing a fuzzy quadrilateral in logical relationship, but gains the maximum space. The entire building has been separated into three parts. There is a public space at the part that facing to the backyard. At this public space, there is a tea house at the first floor, a library at the second floor. There is also a small triangular platform at the library, which has wrapped by the existing trees, merging the trees and building into a single whole. While at the opposite side of backyard, there is set a lounge, book room and the support service space and other relatively intimate space. Between the public space and private space, there is a funning connection space.

The connection space is shaped through a reversing the non-linear hexahedron and then cut off the quadrilateral on its both sides. The connection space merges the different functions at both sides. The stair space has addressed the vertical traffic problem and offered a interior small courtyard to the library for viewing the trees. The connection space has also put new feeling on the basic functional space. The tea house at the first floor has changed its space form flat to vertical. The library at the second floor has also gained a unique sense for the existing of this interesting space.

The connection space is a three-dimensional shape through plain drawing. We have finished the basic analysis and the grasping of space in Rhino. However such difficult model is difficult to change into the construction directly. At the same time, the workers' construction conditions limit designer to provide the measures to realize combining the digital design and the domestic

low-skilled situation together. Firstly, we extract the various skeleton line from the surface in digital soft and make the winding surface be expressed through the interlocking lines. Then, put the straight line divided into equal portions to realize the combination of virtual. The length of the divided portion is controlled within the fitting size of wood template. Then the digital sample has been converted into the situation for manual control. In accordance with this straight line's fitting relationship, they work out the wooden selection mold. On the basic of skeleton, they make the wood plate mold and complete the entire space's template architecture. The template framework is formed in upper and down two times build in accordance with the construction process. The pouring of the plate is same with the ordinary concrete slab mostly. The only one is that the laying of the steels takes the same direction with the plate. The concrete pouring is finished by hand and the entity effect is realized finally. The mold's traces are reserved after construction. The full manual mode of construction made the concrete surface expose a lot of bubbles, the unglued of the template, the steel wire exposed and other quality defects. However, due to the unique form of the winding surface, these defects have been lighten. Though there are defects of the template layout or the pouring process, the combination of digital design and the low skills has endowed special meaning to this digital building.

上海卓美亚喜玛拉雅酒店

上海卓美亚喜玛拉雅酒店隶属的喜玛拉雅中心是上海一座新兴的艺术文化中心，囊括了1100座的大观舞台、喜玛拉雅美术馆、大型购物商场以及5000 m²的屋顶花园。该中心由曾主持设计西班牙巴塞罗那体育馆和洛杉矶现代美术馆的国际著名建筑大师矶崎新先生设计，酒店绝美的内部设计则由曾成功设计卓美亚集团另一地标建筑迪拜帆船酒店的KCA国际倾力打造。其创新的设计理念及非凡的建筑风格灵感来源于中国文化理念和风水原理。

三维立体的异形林犹如参天大树破土而出，诠释了喜玛拉雅中心所要传递的精髓，顶部5000 m²的空中花园连接两座大楼。灵感源于古代树林的根部，融入了风水理念，象征大自然天与地紧密结合。异形林不只作为造型更具备承载空间的结构功能，酒店另一出入口

项目地点：上海浦东新区梅花路1108号
设计单位：矶崎新工作室
设计师：矶崎新
建筑面积：**164549 m²**
主要材料：混凝土,挤塑聚苯板,铝合金玻璃
摄影师：朱少慈

也位处异型林之中。

异形林的建筑设计在中国建筑史上没有任何先例可以借鉴。利用真实树干为模本，以内铸钢丝网外附木模的形式，每隔30 cm用电脑进行力学测算，与自然界树木的生长方式一致，并且保证了工程结构的安全。这项工程没有任何标准进行验收，为此，上海市建委单独设立了验收标准，填补了中国异型施工验收标准的空白。

喜玛拉雅中心的两幢大楼由7层楼高的文字字符包围，由黄帝时期的"造字圣人"仓颉创造。它是对古老的文字和中国文化历史的抽象演绎，也是对当代科技的礼赞。每片文字通过特别定制，并且选能工巧匠细心拼贴，矗立出一道充满想象的"文字墙"。底层的橱窗，巧妙的安排间隔，日光由字符间隔内撒进酒店内部。

酒店楼体结构源自中国古代崇天敬地的礼器"玉琮"形制，内圆外方象征天与地，柱体中贯穿孔高14个楼层，传达天与地之间的沟通，蕴含中国"天圆地方"的宇宙观，量体安排虚实相应，大度恢宏。柱体底部幽静私秘的花园是修身养息的绝佳选择。

酒店大堂内围镶嵌由唐代书法大师怀素书写的千字文。《千字文》最早应梁武帝要求创作，由250个四言短句组成，千字长诗，首尾连贯，音韵谐美，内容有条不紊的介绍了天文、自然、修身养性、人伦道德、地理、历史、农耕、祭祀、园艺、饮食起居等各个方面。

Shanghai Zhuomeiya Himalayas Hotel

Shanghai Zhuomeiya Himalayas Hotel affiliated to Himalayas Center is a new art culture center of Shanghai, including 1100 seats stage, Himalayas Art Pavillion, large shopping center as well as 5000 square meters roof garden. This center is designed by famous architecture designer Mr. Ji Qixin who has ever designed Span Barcelona Stadium and Los Angeles Modern Art Pavillion. The excellent interior design of this hotel is designed by KCA International which has designed another landmark Dupai Sailing Hotel of Zhuomeiya Goup. Its innovation design concept as well as the excellent architecture style comes from Chinese culture concept and Fengshui theory. The three-dimension style seems like a huge tree coming out from the earth, demonstrating the soul of Himalayas Center. The 5000 square meters roof garden has connected two buildings. This inspiration comes from the root of ancient forest, merging with Fengshui theory, which is a sign of combining the god and earth. This unique forest is not only used to as shape and also bear the spacial structure function. The other entrance of Zhuomeiya Himalayas Hotel is also set in this unique forest.

The architecture design of this unique forest has no example to copy in Chinese architecture history. Taking teal tree trunk as the mode, adopting the interior steel net as the form, they are calculated by computer every 30 cm. It has the same grow methods with the nature trees and guarantee the safe of the construction. This construction has no acceptance standards. Therefore, Shanghai Municipal Construction Committee established a special acceptance standards for it, which has made up the blank of China's strange building acceptance standards.

The two buildings of Himalayas Center are surrounded by 7-floor high characters, which are created by Cangjie, "Character Maker" in Premier Huangdi. It is also a kind of praise for modern science and technology. Each piece of character are made by special. After jointing them together carefully, they have formed a "Character Wall" which is full of imagination. The windows of the first

floor allows sunshine coming into the hotel through these character gaps after carefully design.

Zhuomeiya Himalayas Hotel building structure comes from the shape of "Jade Yuzong" which is a piece of ware for worshiping god in ancient China. Round inner and square outer shape symbols of the sky and earth. The hole that passes through the column is 14 floors high, conveying the communication between the god and earth. It contains the universe value of Chinese "Round Sky and Square Earth". They are set in reasonable way, seemingly magnificent. The quiet and peach park at the bottom is a best choice for rest.

The inner room wall of the hall is embedded with Tang Dynasty master calligrapher Huai Su's "Thousand-character Article". "Thousand-character Article" was created in the request of Emperor Liang. It consisted with 250 four-sentence poem and thousands of characters long poems. They were coherent and orderly with harmonic rhythm, which had introduced the astronomy, natural, self-cultivation, human relations, ethics, geography, history, farming, worship, gardening, daily life and so on.

华山路 Chowhaus 餐厅

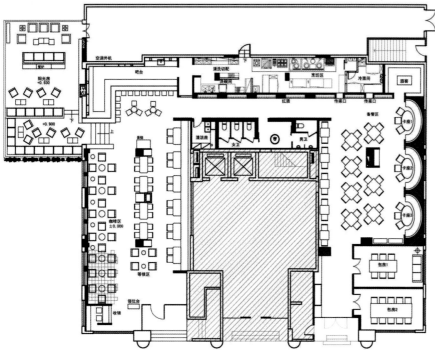

平面布置图
PL-01 SCALE 1:150

项目地点：上海
设计单位：穆哈地设计咨询(上海)有限公司\MRT DESIGN
设计师：颜呈勋 Bill Yen
项目时间：2011
建筑面积：600 m²
摄影师：MOSEMAN ELEANOR ELIZABETH

Chowhaus周边有绿树围绕,门面并不大,木质的外观显得自然低调。

餐厅的开放空间被分成4个区,右边与中间是适合午餐的座位,左侧沙发、小圆桌适合小酌,一般情况下这就是餐厅开放区的全部,可Chowhaus别有洞天。最左往里走,是玻璃房,除却放了不少植物外,中间有个装了壁炉的书架,从各地搜回的老皮箱、旅游手册成为摆设,原色木几周围放着米色、灰色的沙发,自然光线充足,看起来更像个独立的咖啡室。

三个从屋顶悬下的巨型玻璃罩其实是音箱,也就是说,在这件玻璃屋内,你可以带自己的音乐来就餐,营造属于自己的小空间而彼此不会打扰到。相对正式的晚餐区,色调是黑、略深的木色,以及少许金色。内里的两间包房,分别用白色与深木色为装饰,给人以不同色调的冲击感。

Chowhaus at Huashan Rd.

Chowhaus is surrounded by green trees and occupies a small area. Its wooden appearance looks like natural and low-key.

The open space of a restaurant is generally divided into four areas, where the right and middle parts are set with seats suitable for lunch while sofas and small round tables at the left side are arranged for drinking; generally speaking, that is the arrangement of the open space in a restaurant. However, Chowhaus is quite different. Walking inside along the left side, you will see a glass house. There lie many plants and a fireplace-installed bookshelf where old leather trunks and tourist handbooks collected from various places are displayed. The wooden table in primary color is surrounded by sofa in cream and grey colors. With

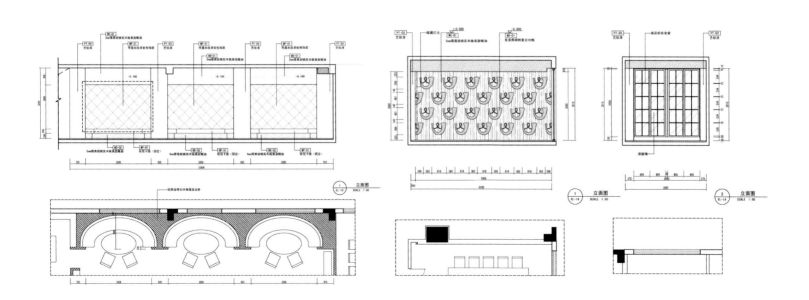

sufficient natural light, the house looks more like an independent coffee room.

Three huge glass covers suspended under the roof are actually sound boxes. So, in this glass house, you can bring music you like to have meals, with your own space made here and without disturbance to each other. The dinner area is relatively formal, where black color and a little dark wood color together with some golden color are designed as the hue. Two compartments inside are decorated respectively with white color and dark wood color, bringing impact sense of different hues.

烧肉达人日式烧肉店/

本案位于上海天钥桥路上,这个年轻的烧肉品牌成立于2007年,是烧肉达人在上海的第三家分店。品牌创立人期望能将日本禅意与中国江南水乡的概念移植到上海,让宾客在舒适优雅的空间里享用美食同时感受到文化的氛围。设计师运用现代的手法演绎日本传统建筑的基本框架结构,大量的木框架朴实的表现建筑结构美学,另外用水墨方式呈现江南水乡中国建筑屋脊的曲线,曲线来自屋瓦依着梁架迭层的加高,藉此强调了这曲线之美在中国建筑结构上几乎不可置信的简单和自然。

项目地点:上海
设计单位:古鲁奇公司
设计师:利旭恒、赵爽、季雯
客户:烧肉达人集团
建筑面积:**300 m²**
项目时间:**2011.11**
摄影师:孙翔宇

YAKINIKU MASTER
Japanese barbecue restaurant

华东 REGION EASTERN CHINA | 117

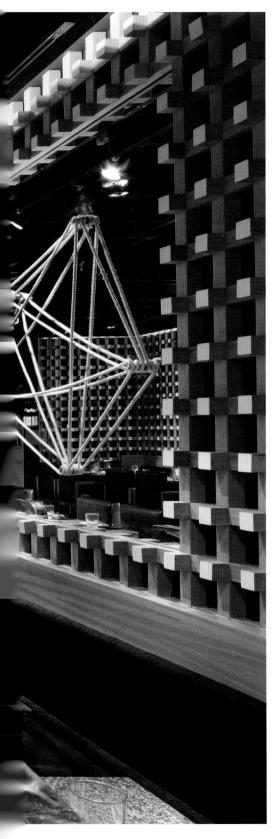

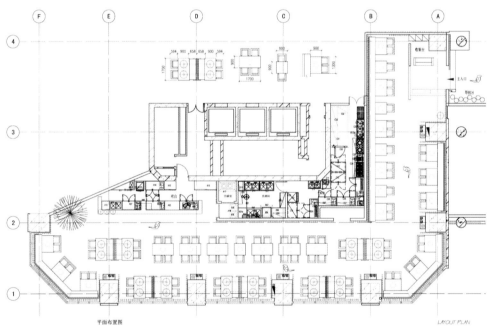

平面布置图　　　　　　　　　　　　　　　　　　　　LAYOUT PLAN

YAKINIKU MASTER Japanese barbecue restaurant in Shanghai Tianyaoqiao road, this young brand, founded in 2007, this project is the third branch of the YAKINIKU MASTER in Shanghai. Brand owners aspire to Japanese Zen transplantation with the concept of south China's country style to Shanghai, so that guests enjoy the meal in a comfortable and elegant space at the same time feel the atmosphere of culture. Designer Lee Hsuheng use of modern techniques to interpretation of the basic frame structure of traditional Japanese architecture, a large number of simple wooden framework of the performance aesthetics of building structures, the other showing the curve of the south of Chinese building roof with ink, the curve from the roof tiles in accordance with the beam the rack layer heightening, thereby emphasizing the beauty of this curve in the Chinese architecture construction almost incredible simple and natural.

上海爱莎金煦全套房酒店

项目地点：陕西北路288号
设计单位：DARIEL STUDIO
设计师：Thomas DARIEL
项目负责人：侯胤杰
摄影师：Derryck Menere

本案的设计理念是将时尚元素带进商务酒店，改变商务酒店一成不变的固定模式，更能符合这个黄金地段定位，与周边的时尚奢侈品牌相得益彰。设计师的灵感源于法国知名西装品牌DORMEUIL(多美)，他将细致的西装布料的条纹元素转化为室内设计的一部分，用艺术感的条纹壁纸或地毯装点空间，即与商务人士相契合，又与时尚不谋而合，更带给商务酒店房间一丝新鲜感。同时，设计师并没有忽略作为一家商务酒店所需具备的功能性，而是将其功能与美观完美地结合在一起。

善于运用颜色的Thomas Dariel在这次的设计中也不可避免表现其对色彩的把握，他在不同的客房里无规则的运用了6种不同的颜色主题：粉色、绿色、蓝色、紫色、灰色、黑色。他的设计初衷是希望带给商务客不同以往的感受，当客人到达时，酒店可以根据客人当时的心情来提供相应颜色的房间，也为客人的再次来访做出铺垫，直到他确定自己最喜欢的房间主题，也可谓带动酒店的再次消费。

对于细节也相当讲究的Thomas Dariel，更是不遗余力的在很多地方展现其追求完美品质的特点。浴室的法国大理石；复古感觉的木地板；开衣柜的皮质小拉手；衣柜门上的优雅的法式复古线型；放大空间感的镜子，让你感觉仍身处奢侈品店里购物试衣。

热爱在设计中隐藏一些小惊喜的Thomas希望由客人自己去发掘其中透路的DARIEL式的幽默。即使在这次的商务酒店设计中也不例外，当你打开衣柜的门时，殊不知竟然只是一扇挡在墙前的假柜门，而门后是一副艺术作品，海报中的外国人正举起相机对着你。装饰的石膏像上带着耳机，只要你把iphone放在前面的坐充上，你就能带上耳机在房间任何一个角落享受音乐的环绕。

上海爱莎金煦全套房酒店经由Dariel Studio的设计改造，重新赋予这幢大楼新的活力并开启其位于这黄金地段的新角色。酒店不仅舒适优雅，设施齐全，而且坐拥市中心的繁华夜景，充分调动五官感受，使出差旅行变成一种享受。

Golden Tulip Ashar Suites / Shanghai Central

The idea of this project is to incorporate fashionable elements in order to alter people's initial impression of a formal business hotel and respect the design of the surrounding luxurious area. The designer uses an eco-striped pattern from good-quality suit fabric, composed by the well-known French brand DORMEUIL, as the wallpaper and carpet in order to create an artistic and modern living space. This style not only brings out a fresh feeling but also illustrates an image of business in a fashionable way. Therefore, the designer combines the aspects of functionality and aesthetics together flawlessly without disregarding the hotel's original intention of being a business hotel.

Thomas Dariel expresses and executes his talented skills by working with six different color themes in this project. To design the various rooms, he combines two colors from a palette of pink, green, blue, purple, grey and black. His original idea of different color themes is to give the guest an unusual experience of allowing her to choose a color theme according to her current mood. This design encourages guests to visit again, inspiring their hopes of finding the room of their choice.

Thomas Dariel effortlessly demonstrates his pursuit in designing with top quality detail. The French marble in bathroom; the vintage wood floor; the elegant French line on the closet door; the mirror for decoration, and the enlarged space all emulate the design of an elaborate boutique store.

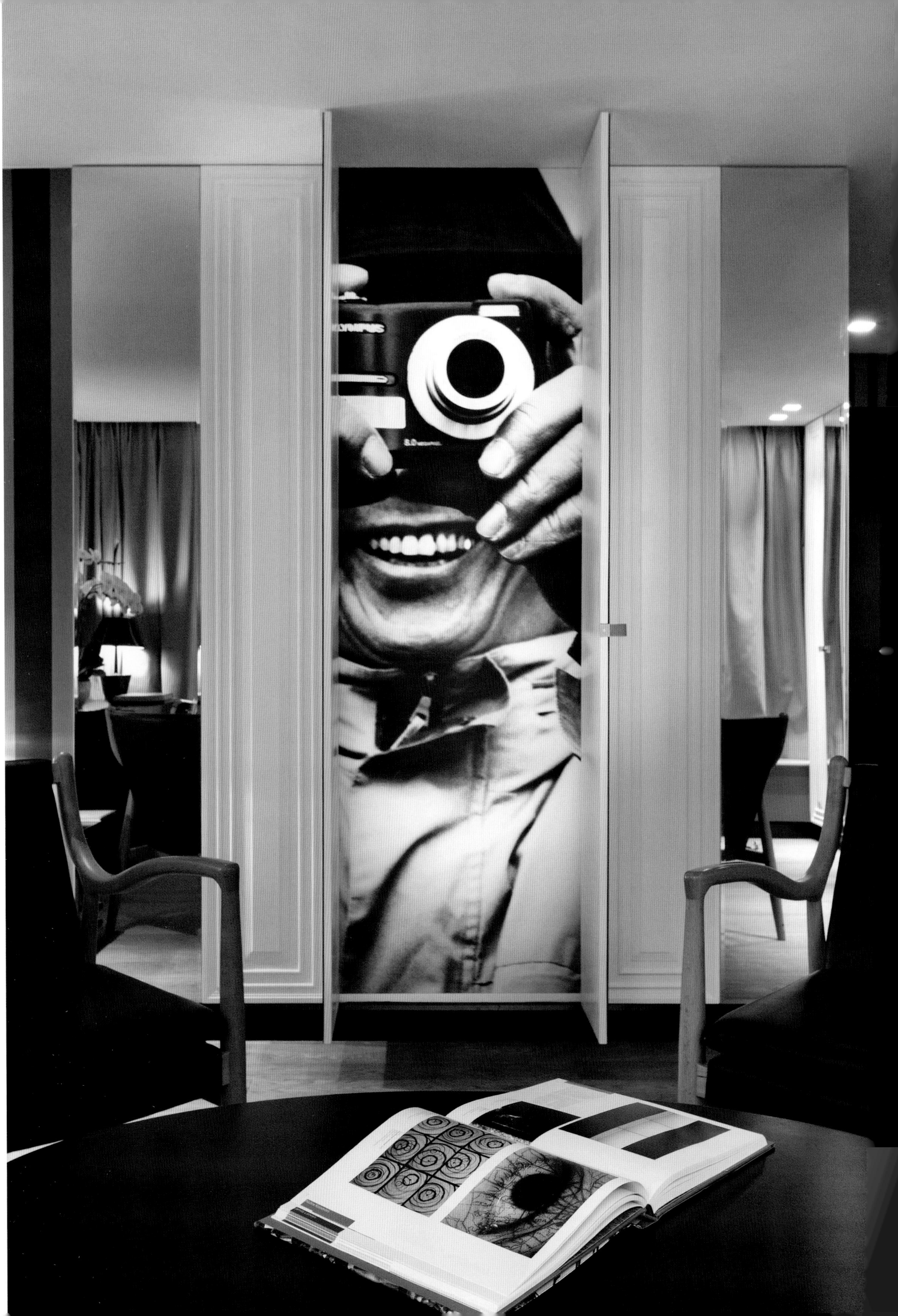
OLYMPUS

Dariel Studio loves hiding some surprises in his design, where guests can discover his unique sense of humour. In the Golden Tulip Hotel for instance, when guests intend to open the closet door, they unexpectedly find a large poster of a man holding a camera in a way as if he was taking their picture. Another surprising, and almost playful touch, is that the door is simply a replica of a closet door. In addition, a head made of deco plaster wears headphones a guest can borrow if he or she wishes to listen to music from their iphone.

After being designed and renovated by Dariel Studio, this building has restored its vitality and is repositioned as Golden Tulip Ashar Suites Shanghai Central. Featuring an elegant and comfortable style, fully equipped facilities, and a beautiful view of Shanghai's bustling city, Golden Tulip Ashar Suites Shanghai Central is a grand hotel that transforms one's overall business trip experience into a pleasurable one.

南京金地名京

项目地点：南京
设计单位：DOLONG设计
施工单位：大品专业施工
主要材料：黑镜、橡木擦黑木饰面、进口墙纸、火烧板、艺术花格、青竹
建筑面积：208 m²
项目时间：2011.05
摄影单位：金啸文空间摄影

"新东方主义"风格，将传统生活的审美意境与现代生活方式有机结合在一起，使家更富有魅力。走进这个居室之中，纯粹的两色既呈现了传统中式的幽幽意境，又无处不契合着居住者所向往的低调奢华，温馨典雅，黑白对比所塑造出的空间灵魂远远超乎了我们的想象。

精装修的房屋总是给人一种冷冰冰毫无生机的感觉，所以业主要求在不改变空间格局的基础上，按照自己对生活的理解，打造出充满蓬勃生命力的居家空间。客厅黑色的"L"型沙发，简洁的线条，超宽的坐面，厚重的色彩，赋予了家沉稳安静的基调，也便搭配各类软装。放置于沙发区中央的浅咖色与米色拼接的地毯，柔软的质地，将家居环境和表达的意境连接在一起，起到了过渡的作用。一幅具有艺术气息的水墨画，被安放在沙发后的墙面之上，在灯光的映衬下，流露出一份中式的深远意境。客厅区域的空间较大，屋主便在这里安排了一个吧台，黑色的吧台，与墙面长方形的字画，搭配的相得益彰，富有禅意的陶罐，使得写意、雅趣的气氛自然散发开来。无论是墙上的水墨画，还是摆放在吧台之上的陶器，这些都是屋主去上海艺博会，为新家精心挑选的，可见屋主是对生活有着颇高的品味。

客厅旁的餐厅，让人感受到另一番景象，餐厅黑色镜面的墙壁传达出强烈的时尚感和沉稳感，与客厅的颜色相协调。为了配合整体的家居环境，主人特意搭配了简约线条的黑色餐桌，亚麻布料的浅灰色餐椅则打破了空间黑色调带来的沉闷感，使房间增添了许多生动的韵律。

穿过长长的走道，便来到了书房，充满着东方韵味的中国元素散落于这里的各个角落。书房中选择了线条简单的棕色家具，以富有中式情结的小饰品做点缀的搭配方式，兼顾了美观和实用的双重标准。书柜上摆放了一幅颇有意境的水墨画、一扇缩小版的中式窗棂、一尊迷你版的佛像……让这个简约的空间

既继承了中国文化凝于方寸空间的博大精深，又呈现出一种清新不失沉稳的格调。

与书房相连的休闲阳台，这里与书房的格调完全不同，我们在这里看见了不一样的景色。这是屋主请做景观设计的朋友精心设计而成，休闲的同时也可以作为会客室，在这里喝茶，看书，聊天，都将是一个不错的选择。石柱上站着的两个小沙弥，手捧着蜡烛，隐约可见的脸庞，为这个空间带来几许神秘的色彩。

家里有个地方，能让疲惫的身体以最快的速度恢复安宁，那就是卧室。房间是宁静的，沉稳的色调，柔软的卧床，一切都给人以舒服的感觉。屋主选择了一款具有东南亚风情的卧床，搭配一款古朴的床头灯，看似漫不经心的安排布置，却让空间闪现出异域的光彩。

Nanjing Golden Palace

"New Orientalism" style, merging the traditinal life aesthetic and modern lifestyle together, make the homoe more attractive.Entering this house, the pure two kinds of color are showing the Chinese atmosphere, highly responding to the low-key luxury, warmly and classical which is loved by people mcuh. The space soul shaped by black and white has far surpassed our imagination.

The refined decoration house always give the feeling of cold feeling, therefore, according to his own understanding of life, the owner request to create a vigorous space without changing the spatial patter.The black "L" shape sofe in living room, coupled with various of soft installation, has simple lines, wide sitting surface, dark color, given this house a calm and quiet feeling. The stitching carpets of light coffee color and the beige color are placed in the center of the sofa area. They connect the house environment and the expression mood together, playing an transition role. An artistic sense Chinese brush drawing is placed on the wall behind the sofa. Under the refection of light, they are expressing a meaningful sense of Chinese style. The living room has a large space. The owner set a bar here. Such black bar coupled with the rectangular calligraphy and painting seem perfect and nice. The pottery here has strengtheded the atmosphere of elegance of nature. No matter the Chinese brush drawing, or the pottery on top of the bar, are all carefully selected by the owner in Shanghai Art Fair. There, we can see a high taste of the owner to life.

Next to the living room is the dinning area, which will give you another experience. The black mirror on the wall has conveyed a strong sense of fashion and calm, in harmony with the color of the living room. In order to coordinate with the whole home environment, the owner specifically select this balck table with simple lines, and light gray dinning chairs with linen fabric. They have broken the dull sense from black and made the room full of vivid elements.

Passing through the long veranda, then you come into a study room, which is full of charming oriental Chinese elements scattered in every corner. This study room selects brown furniture with simple lines. Taking the traditional Chinese tiny products as the decoration, it has considered both beautiful and practical standards. On the bookshelf, there is a piece of meaningful Chinese brush drawing, a narrow version of Chinese window lattice, and a mini version of the Buddha...They have made this simple space inherit the profound Chinese culture that condensate Chinese culture in a limited space. In addition, they are showing a fresh but calm tone.

Connected with study room is a leisure balcony. It is totally different with the study room style. We can see a different scenery here. It is designed by the scenery desiners who are invited by the owner particularly. It can be considered as both leisure area and reception room. Drinking tea, reading, and chatting here all will be a good choice. There are standing two monks on the pillar, holding candles, and looming face, which have added some mysteries to this space.

The place that can relax our body and quiet our heat fast is the bedroom. This bedroom is quiet, stable with soft bed, which have give us a comforable feeling. The owner has selected a Southeast Asian style bed carefully. Coupled with a rustic bedside lamp, even casually arrangement, it has let the space shine the splendor of exotic.

南京紫轩餐饮会所

项目地点：南京市紫峰大厦
设计单位：江苏省海岳酒店设计顾问有限公司
设计师：姜湘岳
建筑面积：1550 m²
主要材料：黑高光木饰面、丹麦灰镜、丝光布、意大利黑金花石材等
摄影师：潘宇峰

紫轩会所，位于江苏第一高楼紫峰大厦4层，定位于南京餐饮业奇葩、绿地广场的美食地标，倾情服务精英人士。

基于这一定位理念，所以从设计之初我们就抛开了非中即欧的传统思想，而是采取了一种融合现代主义和新古典主义风格的设计创想，力求让人们沉浸在优雅的文化氛围中亦能品味到现代时尚气息，即古典式现代美学之设计理念。

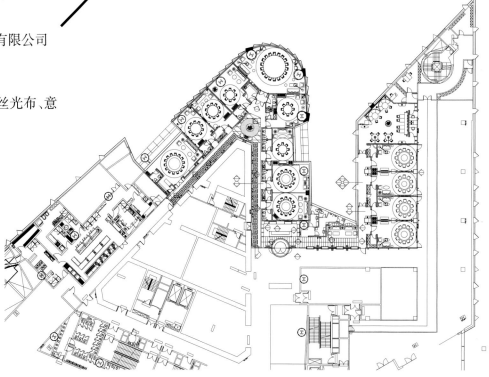

Zixuan Club, is located at the first building of Jiangsu Province---4th floor of Zifeng Building, which is positioned as the maracle one of Nanjing restaurant industry, garment landmark of green square, servicing talents by fully heart.

Basic as this concept, therefore, at the initial design part, we has abandon the Chinese style or the Europe style. By contrast, they adopt a design concept merging modernism and neo-classical style. It is strive to let people immersed in the elegant atmosphere and enjoy the modern and fashion sense. That's the classical modern aesthetic design concept.

Nanjing Zixuan Restaurant Club

江苏亚明室内建筑设计有限公司办公室

项目地点：南京市1865创意产业园
设计单位：江苏亚明室内建筑设计有限公司
设计师：孙亚明 许良明
建筑面积：700 m²
主要材料：铝板、优洁环保板、大西洋灰石材、不锈钢、地砖、夹胶玻璃
项目时间：2011.03
摄影师：文宗博

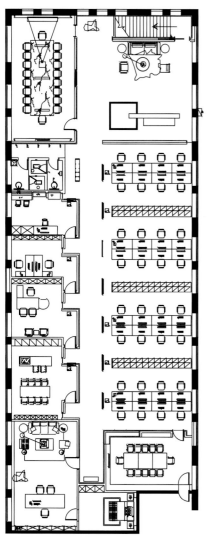

简约的设计理念、超常规比例和尺度,塑造出强大视觉感。体块和光影的变化,强调面与面穿插,增大了空间尺度。空间主色为白色,以绿色景物衬托,使人觉得清新自然。水景、鱼缸等元素的注入,让人身处其中,感受到自然的气息。整个设计用材采用了系统集成装饰手法,便于拆装、二次使用,真正做到了绿色环保意念!

Jiangsu Yaming
Interior Architecture
Office

Simple design concept, super normal proportion and scale, they have created a strong visual sense. The changes of body and lighting has highlighted the interspersed between surfaces, letting people feel fresh and natural. It also has enlarged the space. The space takes white as the main tone, taking green plants as supplement. Water, fish tank elements have given people a true feeling seemingly like stay in the nature. The whole design adopts combination methods which is convienent for dismantle and second time usage. It has achiebed the green and environment friendly concept.

瓦库 6 号

"瓦库"以众多瓦的集结,用"瓦"单纯的合声呼唤我们记忆的情感,借以为都市人们提供一个喝茶叙旧的地方。

打开每扇窗,阳光照进,空气流通是瓦库设计的核心理念。空间布局不为风格而风格,一切为阳光和空气让路,瓦和吊扇承担着主角的使命,空间自然而就。

"瓦"所唤起的文化记忆,"阳光、空气"给人们带来的好处,使它赢得了经济效益与社会影响的双丰收。

项目地点:南京奥体
设计师:余平
参与设计:马喆、董静
建筑面积:**1000 m²**
主要材料:红陶瓦、土青瓦、土陶瓦、户外地砖、旧实木、瓦脊构件、古民居图片、土陶罐
摄影师:贾方

Waku No.6

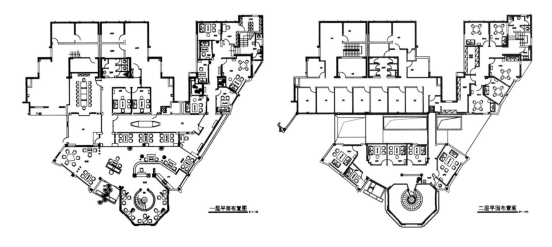

一层平面布置图

二层平面布置图

"Waku" is a collection of tiles. Using "Wa(tile)" to collect our memories, it has provided a tea place for urban people.

Open each window, let the sunshine pouring in, and flowing air is the core concept of it. The layout is not pursuiting only for style but to let sunshine and fresh air in. Tiles and the suspended fan bear the major mission. The space is designed naturally.

Waku No.6 was opened in October, 2012. "Wa(tile)" has collected all cultural memories. "Sunshine and air" have benefited people to people, which has allowed it gained both economic efficient and social influnce.

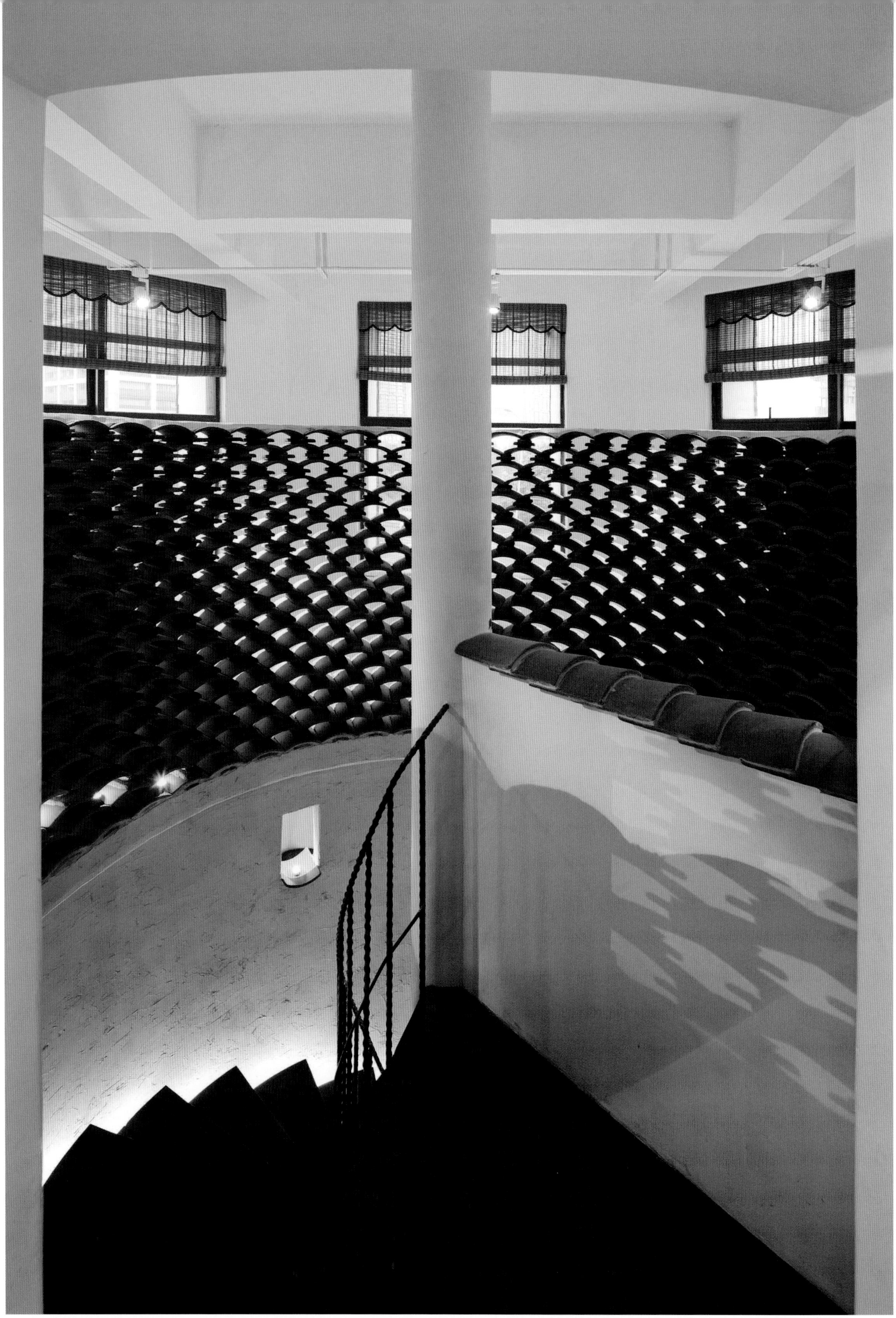

万濠华府会所

项目地点：江苏南通
设计单位：KLID达观国际设计事务所
设计师：凌子达
建筑面积：3800 m²
主要材料：咖啡洞石、闪电米黄、黑金砂、灰木纹、黑檀木

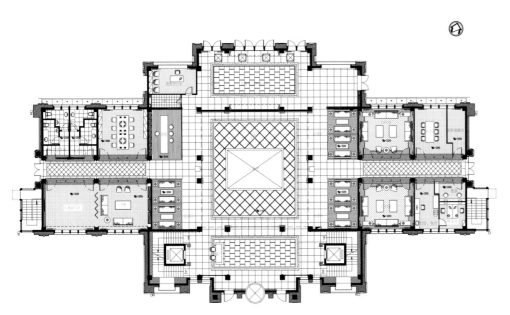

本案为住宅项目里的一个会所,会所将来保留给住宅业主使用,是一有多功能用途的会所。

设计风格为简约的古典主义,运用大量的咖啡洞石、闪电米黄与黑檀木塑造出一种奢华感。在立面设计上大量并重复地运用了"八角形"的形式,重塑了ArtDeco风格。

Wanhao Huafu Club

This case is a club of the residential projects. The club will be reserved for the future owner. It is a multi-function club.

Its design style is simple and classical, using large amount coffee travertine. The shining beige and black ebony are trying to generate a sense of luxury. For the facade aspect, the designer uses many octagonal form repeatedly, wearing an Art-Deco style on it.

星光捌号

项目地点：无锡
设计单位：PANORAMA 香港泛納设计师事务所
设计师：潘鸿彬、谢健生、蔡智娟
建筑面积：700 m²
主要材料：紫红色光纤、枕木水泥壁、生铁、黑白牛图案
项目时间：2012
摄影师：吴潇峰

星光捌号是无锡市中心新开设的一间牛排餐厅，位于历史活化项目"西水东"的百年工厂房之内。设计概念是将富有历史价值的仓库加以改造更新，保留其大部分建筑的原有风貌。加建的夹层结构大大的增加了可用空间，使它成为时尚浪漫的现代化餐饮地标。设计策略是将传统的牛排屋餐饮体验提升到一种新的境界。餐厅的两个用餐区内多种类型的座位的布局迎合不同顾客的要求及在这时尚而浪漫的意境中享用锯扒之乐。

中庭用餐区保留原有旧建筑的元素，包括10 m高屋顶的斜面天窗和梁柱结构、外墙红砖和室内水泥壁及窗框等，使顾客可以一边进食、一边欣赏老建筑的新面貌。2 m高的雄牛屏用刀叉组成，他是餐厅的品牌标志，使顾客一见顿生食欲，区内垂直宽敞的空间里，6 m高镜钢铁酒柜和VIP区顶上的水晶吊灯是装饰重点，中央和周围用餐区着真皮梳化，天花垂吊著紫红色管线造成的点点星光，缀成了星光下的晚宴。

夹层用餐区新建的钢构夹层是半开放式。其多功能区的黑白牛图案吊顶和地灯使顾客到此有"回家"及宾至如归之感。半透明的活动屏风装置提供活动的间隔，以便满足举行不同的活动需要。最后，顾客在黑镜饰面的洗手间里结束了其星光之旅，留下了难以忘怀的印象。

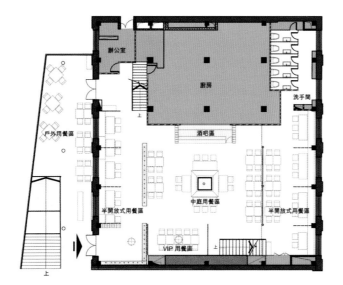
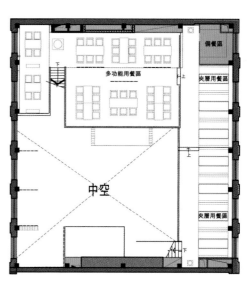

"Starry Night Dining"

StarEight is a new steak house situated at one of the warehouse buildings within a 100-year old factory compound in the city of WuXi, China. A historical re-vitalization design strategy was adopted to preserve most of the original building features with addition of new mezzanine structure to maximise the spatial potentials and turned the old warehouse into a hip F&B landmark. A main scene of "Starry Night Dining" was created to move the traditional Steak House dining experience to a new level of trendiness and romance. Various seating patterns were created in the two dining zones to cater for different customer needs:

Atrium Dining Zone – Original building

features of the10m high pitched roof skylight, truss structure, exterior brick & interior fare-faced cement walls, window frames were well-maintained and exposed to the eyes of the customers. 2m high bull-shaped partition made by knifes & forks gave clue and defined brand identity tothe restaurant. Verticality & spaciousness of the atrium dining zone were emphasized through full-height mirror stainless steel wine rack and crystal chandelier above the VIP table. Ceiling suspended fibre optics in violet colour created starry night effect to the leather-upholstered central & periphery booth seating areas.

Mezzanine Dining Zone - New built steel structure mezzanine floor was introduced to provide a semi-enclosed dining zone. VIP area with black & white bull-patterned ceiling-wall and floor lamps gave cosy & home feelings to the space, sliding semi-transparent curtains provided flexible partitioning for holding different events. The romantic starry night experience was finally completed at the black-mirrored washrooms and left an unforgettable memory to the customers.

无锡灵山精舍

无锡灵山精舍坐落于无锡灵山胜境内，毗邻灵山大佛，总体建筑规模9800 m²，拥有90多间客房，都掩映在一片安静的竹林之中。来这里客人们可以安下心来修身养性、体悟禅境并参加精舍里提供的各项与参禅相关的活动。在精舍的室内设计中，设计师以"竹"为母题，传承佛陀千年前在印度竹林精舍时的意境。一入大堂，木质的带有风化感的条形格栅天顶、旧铜打造的前台，还有竹子做的几盏大吊灯让人的心一下子就沉静下来。朴素的客房，简单但也很精巧，透过细密的竹帘，目光可以穿越到窗后禅意的小院子里。禅堂素雅，竹子的顶棚，竹子的天灯都照应着向禅的心灵回归自然无我。茶室里的家具简约但也渗透出禅境，让客人更好地在参茶的过程中调节心境。这所精致Resort正以"禅"为主题，提供客人"禅"的教诲，"禅"的感悟，"禅"的意和境。

项目地点：无锡滨湖区
设计单位：HKGGROUP
设计师：陆嵘、慎曦、李婷、田珺
建筑面积：9800 m²
主要材料：竹、实木、青砖、竹编藤编、墙纸、青灰色凿毛石材、木地板、鹅卵石、图案地毯、透光膜等
摄影师：刘其华

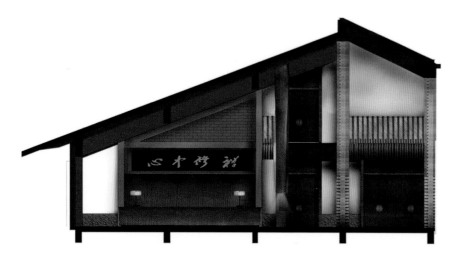

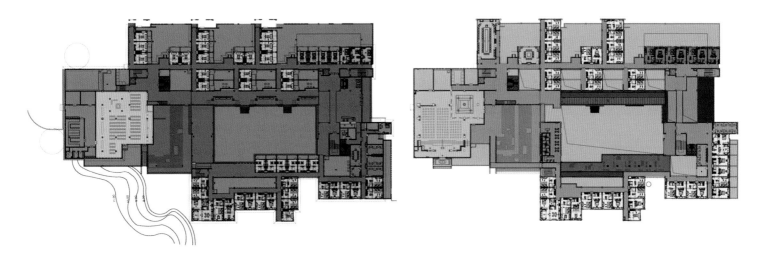

Wuxi Lingshan Jingshe

Wuxi Lingshan Jingshe is located at Lingshan Mountain, Wuxi City, closing to Lingshan Buddha. Its overall construction area is about ten thousand square meters, bodering 90 rooms and hiding in this quiet bamboo trees. The guests here can calm down themselves, enjoy the Buddha spirit and participate various Buddha activities provided by Wuxi Lingshan Jingshe. For the internal design of this Jingshe, the design takes bamboo as the main tone, inheriting the Indian bamboo pagodas happened thousand years ago. The time you entering this hall, the wooden with weathered sense zenith, the old copper reception table, and several chandelier made from bamboo will calm your heart and soul suddenly.

扬州富临壶园府邸

项目地点:扬州广陵区
设计单位:苏州徐晓华室内设计有限公司
设计师:徐晓华
参与设计:苗永光、李爱岭
主要材料:石材、地板、成品木饰面板、布艺硬包、墙纸
建筑面积:**1800 m²**
项目时间:2011.01-2011.08
摄影师:潘宇峰

壶园是一座极具江南特色的私家园林,始建于清代,设计师在其室内设计构思时也结合这一特色,加上现代生活学,给人风雅、舒适的感觉。餐厅地面是深浅纹石材辅以实木拼花地板,深色的圆型柱子加上订制的艺术灯具等复古元素的点缀,整个空间流露出高雅与低调的奢华感。包厢里摆放造型简单、颜色淡雅的家具,复古的陶罐摆设,古典韵味的玉器、贝壳等软装的陈设使一种宁静、高贵、雅致的感觉在整个空间中蔓延。庭院、走廊设计师运用新中式的表现手法,让整个空间又多了一份优雅的文化气息置身其中,让人流连忘返。

Yangzhou Fulin Hu Garden

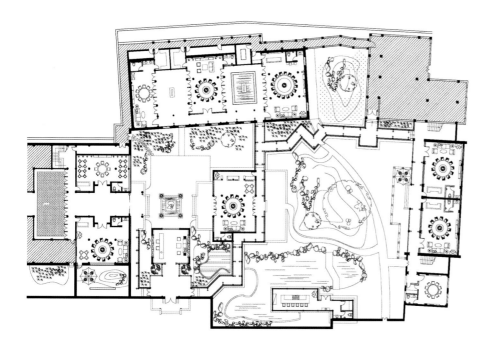

Hu Garden is a most Jiangnan private park. It was built in Qing Dynasty. The designer has merged this character into the space. Coupled with modern life, it gives people a sense of elegant and comfortable. The restaurant ground is made of stone and solid wood parquet floor. The dark round pillars coupled with the customed art lamps and other vintage decorative elements, the entire space seems so elegant and luxury. In the room, there is placed simple elegant furniture, retro pottery and classical jade, shells and so on. They are showing a kind of peace and noble sense, which also extend to the whole space. For the courtyards, corridors, the designer used new Chinese technique to express, letting the space add an elegant cultural atmosphere, making the space full of charming.

华东 REGION EASTERN CHINA | 179

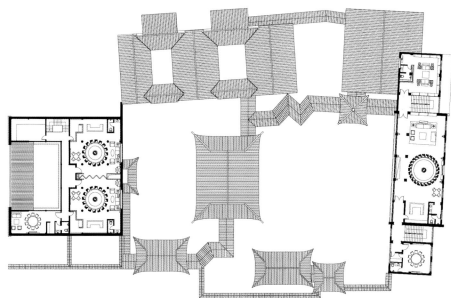

裸心谷

首席规划及建筑单位：benwood Studio Shanghai
夯土小屋建筑单位：A00 Architecture
园境建筑单位：Design Land Collaborative (DLC)
环保：Environmental Resources Management (ERM)
可持续性：安生态有限公司 (BEE)
LEED：上海太平洋能源中心 (SPEC)
室内设计：AIM
室内空间规划：benwood Studio Shanghai

裸心谷位于浙江省风景秀丽的田园胜地莫干山，这个强调可持续理念的豪华养生中心兼牧马自然保护区占地67英亩，坐落于一个私人山谷之中，四周环绕着大型水库、翠竹、茶林以及一些小村庄。度假村内拥有121间客房，分布于独栋树顶别墅以及夯土小屋之中——所有这些建筑全部采用业内领先的可持续材料建造。度假村内的用餐选择包括拥有80个座位，位于水库旁边的中非合璧餐厅Kikaboni（在非洲语中意为"有机"）、泳池酒吧兼西餐厅Clubhouse Cafe和24小时的客房送餐服务。度假村内拥有占地750 m²的养生水疗中心naked Leaf，中心的15间理疗室设于竹林内的高脚屋之中，在竹影婆娑下更显宁静安谧。裸心谷更设有一个1000 m²的会议中心Indaba（在非洲语中意为"首脑集会"），拥有8间多功能会议厅和2个分别俯瞰着竹林和水库的大型平台。此外还设有一间会所和一间名为Little Shoots的儿

童游乐兼托儿中心。

　　Naked Galleries 是针对追求文化内涵的休闲或商务旅客而设，位于会所往上的四栋风格独特的建筑环绕着一个小湖，每栋建筑都拥有特色文化主题：茶亭、竹艺馆、项目馆及陶艺室。

　　树顶别墅和融合亚非风格的室内设计均出自 Benwood Studio Shanghai 的知名建筑师 Delphine Yip-Horsfield 之手。Benwood Studio Shanghai 负责的知名大型项目包括集餐饮和娱乐于一体的上海新天地等。夯土小屋则是由上海本土环保建筑设计师 A00 Architects 操刀。匠心独运的室内外设计，令秀丽的山野景观更显迷人，带来返璞归真的自然体验，同时最小化对环境带来的影响。

　　30栋树顶别墅均采用两层设计，提供豪华的双卧室、三卧室或四卧室套房。高挑的楼层和从地面到天花板的玻璃幕墙带来无障碍的绝佳视野。别墅的第二层高于树顶，室内为设备齐全的厨房和客厅，而开阔的室外露台设有浴缸、烧烤台和用餐区。所有房间和别墅均配有高速无线上网、36寸卫星电视、CD机、DVD机、iPod基座、高级音响和中央空调系统。水疗式淋浴、纯天然卫浴用品、独家床品和随传随到的管家服务，为您送上极致奢华的享受。七栋特别设计的贵宾树顶别墅拥有独一无二的迷人景观，提供更周到的服务和配套设施。

Pure Heart Valley

Pure Heart Valley is located at beautiful Mo-gan Mountain, Zhejiang Province. This luxury health center is pursuing of sustainable development concept, covering 67 hectares. It is also a Wrangler Nature Reserve locating at this private valley, surrounded with huge water pool, bamboos, tea plants and some villages. This hotel possesses 121 rooms, scattering in different independent villas and small ancient houses—all these architecture are all adopted the advanced sustainable materials. The restaurant in this resort possesses Kikaboni (means "organic" in Africa language) with 80 seats that locating beside water pool, Clubhouse Café that has swimming pool, bar and western food, as well as the 24-hour room service, there is a naked Leaf water spa covering 750 square meters. These 15 spa rooms are placed in the bamboos, which seem quieter and peace under the shadow of these bamboos. Pure Heart Valley even has a meeting center Indaba (means "President Conference" in Africa language) covering 1000 square meters. It has 8 functional meeting rooms and two huge platform to enjoy the bamboos and water pools separately. In addition, there is a club and a room named Little Shoots children playing center.

Naked Galleries is set for guests that pursuit of cultural leisure and business function, locating at four unique architectures surrounding a small lake ahead of the club. Each building has its own unique culture theme, Tea House, Bamboo Art Pavilion, Project Room and Porcelain Room. People can try to make the most location handmade arts, tasting the "white tea" generated by Pure Heart Valley, making a tea pot on the porcelain turntable, pick up the fresh tea leave by yourself, experiencing the local life style, bamboo plant as well as the various environment friendly projects cooperated with the local famers. The round theater locating at the center will hold concert and cultural performance.

The villa located higher than trees and the Sino-Africa indoor design both come from the famous architecture Delphine Yip-Horsfield of benwood Studio Shanghai. The famous huge project that charged by benwood Studio Shanghai includes Xintiandi Shanghai that merges restaurant and entertainment into a single whole. The ancient small house is made by the local architecture A00 Architects. The special indoor and outdoor design making the beautiful mountain more charming, bringing us a natural and true experience. They are all environment friendly.

These 30 villas located higher than trees adopt two floors design, providing luxury two-bedroom, three-bedroom or four-bedroom suits for selection. The high building and the wall glass window has brought a perfect vision. The second floor of the villas is higher than the trees top. They are all well-equipped with kitchen and halls, bathtub, barbecue, and dining area at the outdoor terrace. All rooms and villas are all equipped with wireless internet, 36 inches satellite TV, CD, DVD, ipod Base, high-class stereo and central air conditioning system. Spa shower, pure natural washing products, exclusive bedding materials and the on-time service will provide you a luxury experience. Seven special villas possess unique and charming landscape with most perfect services and equipment.

万科良渚文化村少儿会所二期

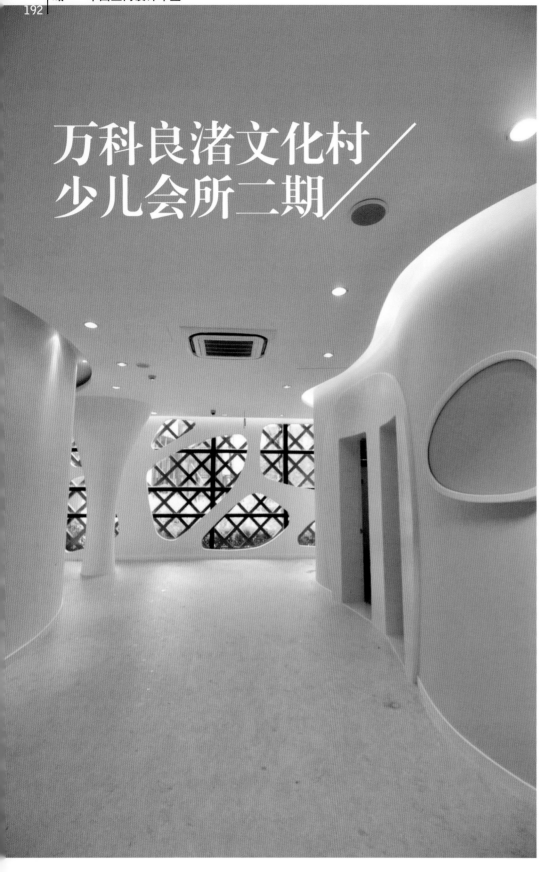

项目地点：杭州余杭
设计单位：杭州海天环境艺术设计有限公司
设计师：姚康荣
建筑面积：1200 m²
主要材料：彩色地胶板、pvc地胶板、石膏板
哑光漆、彩色乳胶漆
竣工时间：2011.01

  水滴的符号是贯穿本案的主题符号，一个个大小不一的水滴嵌在光滑的墙面，犹如美妙的音符欢快的跳跃，营造了一个充满活力的空间。
  良好的幼儿活动场所使幼儿在认知、交流、行为、色彩、艺术等得到良好的熏陶与锻炼，对幼儿的成长至关重要。幼儿园是连结家庭与学校的桥梁，是儿童的乐园、儿童的家。

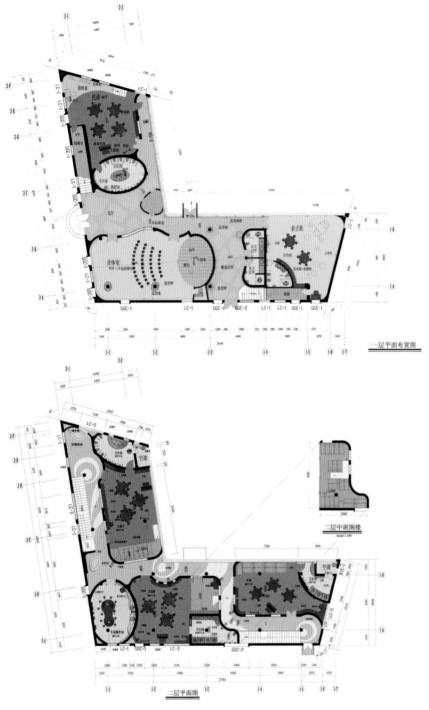

The second phase of Vanke Liangzhu Culture Village Children Club

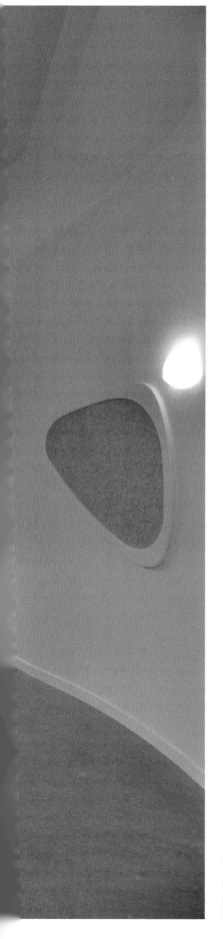

Water drip symbol is the main theme of this case. The water drips are embedded on the smooth walls, seemingly like the jump rhythm, creating a vibrant space.

The well child playing place allow the children get well influence in terms of cognitive, communication, behavior, color, art and so on. It is very important for children's growth. Kindergarten is the bridge linking the family and school, is the paradise of children.

HENG LI FABRIC ART SALES CENTER

项目地点：浙江海宁许村
建筑面积：**380 m²**
设计师：周伟
参与设计：盛汉杰
主要材料：黑洞石、人造石、亚光砖

恒立布业销售中心

室内是建筑的延续，空间感才是建筑的精髓。在一个长25 m，宽10 m，高6 m的空间中做一个销售中心，在造价有限的前提下我们一反常规的工作程序，先做空间，做完空间再把各种功能合理的安排到位。于是出来现在这样的空间效果。

在空间的处理上，我们通过盒子的挑空、坐落、衔接而组成不同的室内空间，从而满足这个空间上的一些功能需求。一部折线楼梯贯穿整个空间，将整个空间连接在一起。入口坐落的是个前台，在黑色石头墙面的映衬下由极几何感的服务台好似一座雕塑。过道上的连廊划出了一道趣味的空间关系，站在挑空的连廊上俯看整个空间似乎又多了几分韵味。挑空的大盒里面是一个办公的地方，里面的一切活动都是围绕着一张大的曲线桌子而展开，在喇叭形的灯光的作用下，在里面办公一定是件极具趣味的事情。而坐在盒子下面的吧台上听着音乐休息之时瞄眼整个空间又有着另一份味道。

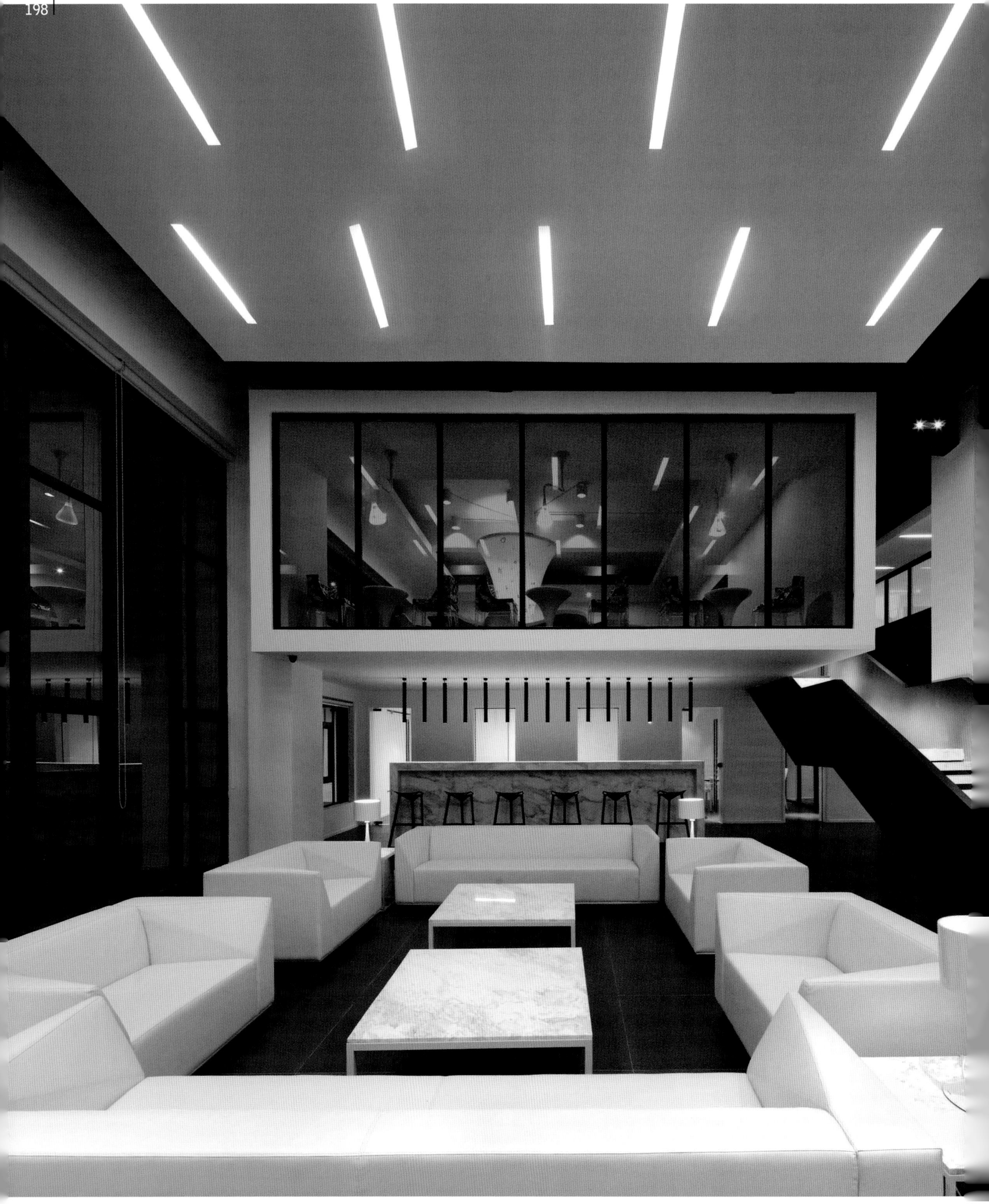

Hengli Cloth Industry Sales Center

Indoor is the extension of the construction, while space decoration is the soul of the construction. In the space of 25 meters long, 10 meters wide, and 6 meters high, within the limited cost, we make a unconventional process to build this sales center. We put space first, and then put a lot of functions in a suitable place. Thus, we get the final space.

On the treatment of the space, we connect the different spaces through hollow, seats and then satisfy the function requirements of the space. A line of stairs has connected all the different spaces. At the entrance, there is a reception table. Under the reflection of black stone wall, this reception table looks like a sculpture. The corridor has drawn an interesting picture in this space. Standing on the corridor, the space seems to have more charming. At the hollow huge box, there is office area. All activities are conducted around this huge winding table. Under the trumpet-like lamp, working in this office area seems full of funny. At the same time, sitting beside the bar, and listening to the music, you will feel another taste of this space.

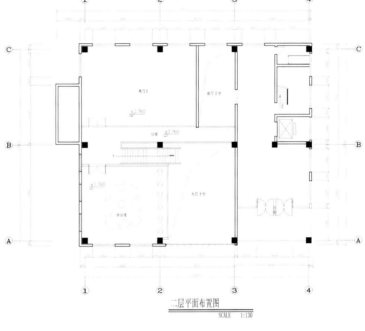

二层平面布置图
SCALE 1:130

一层平面布置图
SCALE 1:130

UTI品牌女装文一店

项目地点：杭州文一路
设计单位：杭州观堂设计
设计师：张健
项目时间：2011.04 – 2011.07
建筑面积：181 m²
主要材料：砖墙、木质
摄影师：王飞

UTI品牌女装，设计上要求清新、自然，整体店铺以白、灰色调为主。设计过程中，着重采用了"解构主义"的手法，比如墙体处理上，留下一些斑驳的痕迹，或开凿出不规则的门洞；收银台方面，将各式抽屉、门板分解后再拼凑；装饰柜方面，采用柜体分解再合成的设计；这些解构处理，既能带来设计上的新鲜感，又能体现品牌面料的解构使用；相比硬装的简洁，后期软装上，则花费了诸多心思。货架体系上，采用木质为主，暗合"清新，自然"的追求，简洁不代表简单直白，在货架货柜设计中，采取了一些欧式线条的处理，简单的木质在大方的线条中，变得柔和温暖；摆设上，选用了诸多东南亚原木、柚木家具，如木墩、椅子、柜子、圆桌、木马等；软饰方面，采用回收的老皮箱、工业时代灯具，显现别致的品味，营造素雅的氛围。同时，设计师延续了其一贯的环保理念，在顶与地的处理上，适当的选用了回收木料，贯穿"再利用"的思想。

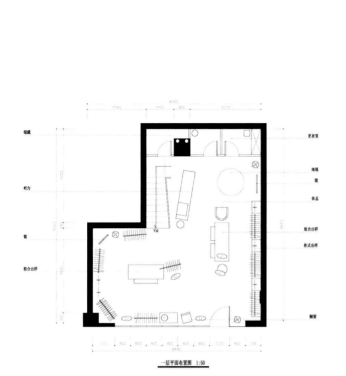
一层平面布置图 1:50

二层平面布置图 1:80

UTI woman's brand, requires clear, natural in design. Its whole shop takes white and gray as its main tone.

During the design process, it is adopted "deconstruction" methods greatly, such as the treatment of wall and leaving some mottled traces, or cut some irregular door holes. On the cashier place, they decomposite the doors and drawers firstly and then put them together again. In terms of decoration cabinet, they still adopted the deconstruction method first and then put them together. These deconstruction method can both bring a fresh design and reflect the full usage of the brand quality fabrics.

UTI brand pursuits of natural and clear style. Therefore, on the shop design, they do not adopted the complex decoration but some basic treatment simply. The cement floor, brick wall with white paint, they are all treated with the same color and same materials and no more spare materials.

Design Concept

Enter this shop, you will feel simple and greatness.

Compared with the simple hard decorations, the designer paid much attention on the soft decorations latter. For the good framework, they are taking wooden as its main material, which implies the pursuit of "clear and natural". Simple is not boring, on the framework design, it is adopted some Europe-style treatment on lines. Then they become more warm in terms of appearance. On the layout, they have selected many Southeast Asia's original wood and teak furniture, such as the wooden pier, chairs, cabinets, round tables, wooden horse. On the soft decorations, it is adopted the recycled old suitcase, the lamps of industrial era, which are showing a unique taste and creating a simple and elegant atmosphere.

Meanwhile, the designer has also continue his environment friendly concept. On the treatment of the ceiling and floor, he has selected recycled wood maters, which reflects the "re-use" concept.

上虞宾馆

项目地点：浙江上虞市
设计单位：杭州大相艺术设计有限公司
设计师：蒋建宇、李水、董元军、郑小华、楼婷
建筑面积：**5.2万 m²**
主要材料：柚木、木化石、珍珠黑、西西里灰
摄影师：贾方

上虞宾馆为江南园林风格度假商务酒店，是一家山景合院式度假商务酒店，本案运用复兴传统风格设计，将传统、地方建筑的基本构筑和形式保持下来，加以强化处理，突出文化特色以及地域特色——上虞特有的"虞舜文化"、"青瓷文化""江南文化"。

酒店项目位于上虞的一个私家山顶上，山体后高前低，自然景观优美且浑然天成。以合院为主题的设计不仅蕴藏了深厚的地域性文化情感，也忽略了酒店本身的商业气息。同样源于这一饱经历史沉淀的设计命题，使建筑体更加自然地相融于自然景观之间。设计上汇集了诸多与中国古典文化及上虞本地历史文化相关的设计元素，凝聚了深厚的人文情感，让历史传承中的四合院精神，在宛若世外桃源的山上重新演绎着早年的风情。情景交融不再是片刻的感慨，进入到建筑的每寸空间都可以呼吸到自然的气息。宽绰明朗的空间，纵观全景的玻璃墙，俏而争春的盆栽，俨然与户外景观浑然一体。以感怀的心去触摸"合院"内心的丰富，以风格独特的建筑室内空间、品味独具的艺术品鉴赏，高调享受生活。体贴入微的酒店设计，大

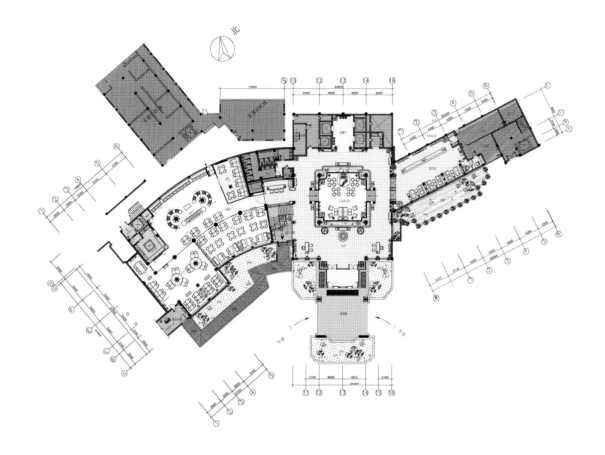

一层总平面图
LEVEL 1 KEY PLAN

堂、西餐厅、餐厅、客房,让身处酒店的客人也能感受到家庭般的温馨。

设计借由材料的运用,巧妙地将中国文化融合其中,使新、旧感受并列且同时呈现出东、西方文化交融的独特风格。汇聚中国古老元素与现代工艺科技的上虞宾馆使用了通常被用在建筑外墙的灰砖作为内部装修建材之,营造出建筑的特殊美感与功能,设计上使用如今少用的传统建材及传统文化的题材作为空间永恒典范的又一构思。在每个空间内部随处可见的木雕屏风与青瓷洗脸台再度验证了设计师刻意交织古今与中西于一体的设计巧思。以现代建筑艺术来阐释中国江南建筑风格的度假商务酒店。

居住之外,接待区、休息区、多功能会议室、精美百态宴会厅、顶层空中花园共同构建宏阔华丽的会所式酒店,营造着贴合人性本质的盈动与大气。整个酒店无论是在布局、摆设和细节处,都展现出厚重丰饶的人文享受,庄重而不显呆板。现代设计手法与古老风格元素相结合的表达方式巧妙地呈现出具有冲突感的视觉效果,却也成功营造出理想化的儒雅情景。

有别于传统豪华酒店所提供的服务,上虞宾馆秉承让客人独享"雕琢奢华"的理念,将度假胜地的感觉巧妙地融入于当代都会空间中,形成低调奢华和内敛雅致的现代触感,现代风格与复古主义相互融合,丰富的感官体验,让宾客沉浸在个人专属奢华所带来的全新感受中。

Shangyu Hotel

Shangyu Hotel is of Jiangnan garden style and for vacation and business. It is the only hotel with mountain scene and courtyard at the bank of Nanhu Lake in Zhejiang. In this case, traditional style is applied to keep basic structures and forms of traditional and local buildings and then strengthen them, thus highlighting cultural and regional characteristics – "Yuhun Culture", "Celadon Culture" and "Jiangnan Culture", unique in Shangyu. The hotel project is located on a private mountain top in Shangyu. With high rear and low front, the mountain owns natural and beautiful scene. With courtyard as theme, the design not only contains rich regional culture emotion but also eliminates the commercial atmosphere in the hotel. The design theme which also originates from the long history makes the building more naturally integrated with the natural scene. The design is full of elements related to Chinese classical culture and Shangyu history and culture and contains rich human emotion. It allows the courtyard spirit inherited along the history to show its early appearance on this Arcadian mountain. Feelings and scenes are mingled together not just for a while; in every space you step into, the nature can be felt. The wide and bright space, glass wall with a panoramic view, and flourishing plants seem to have mingled with the outdoor scenes. Touch the rich inner of the "courtyard" with your emotional heart and enjoy the life as much as you like with the in-

door space of distinctive style and art works of unique taste. The thoughtful hotel design in hall, western-style restaurant, restaurant and rooms makes customers feel at home.

The design can be explained in detail as follows: located at the bank of beautiful Huanghai Sea, elegant and not artificial, honorable and not flamboyant; low-key and implicit, inheritor of Chinese traditional culture, collector of national spirit and strength; with a quite and elegant environment inadvertently making people contemplate and taste and a pleasant atmosphere allowing people calm and content; solemn and steady, generous and refining, meaningful and long lasting, full of ecological feeling and human emotion, making people feel intimate; aiming at classical Chinese style based on detail consideration and overall coordination. By application of materials, the design skillfully blends Chinese culture, bringing both old and new feelings and showing a unique style where Eastern and Western cultures are mingled together. Full of both Chinese ancient elements and modern process technologies, Shangyu Hotel applies grey brick (usually for outer wall) as one of interior decoration materials, to create special building beauty and function. In addition, building materials and cultural theme of traditional style, which are scarcely applied in design, serve as the eternal space model. The carved wooden screen and celadon basin can be seen in every space, which again verifies the designer's excellent idea to mingle the ancient with the modern and the China with the West. The Chinese Jiangnan building style of this resort and business hotel is shown with modern building art. In addition to rooms, reception area, resting area, multi-functional meeting room, fine banquet hall

of various styles, and roof garden make up this grand and gorgeous chamber-type hotel where spirituality and generosity is created to cater for human nature. The solemn and flexible hotel shows plentiful humanistic enjoyment in terms of layout, display and details. By combination of modern design techniques and old elements in the expression way, both conflicting visual effect and ideal genteel scene is created skillfully and successfully. Different from traditional grand hotels in service provided, Shangyu Hotel holds the idea to allow customers to enjoy "luxurious carve" alone. The feeling of a resort is skillfully mingled in modern city space, to form low-key luxurious and elegant modern feeling. The fusion of modern style and restoration together with rich sensory experience immerses customers in a new feeling brought by personal exclusive luxury.

零壹城市事务所办公室

项目地点：浙江省杭州市
设计单位：杭州零壹城市建筑咨询有限公司
设计师：阮昊、李煦、王鹏
项目时间：2011
建筑面积：350 m²
照片提供：零壹城市建筑事务所

零壹城市建筑事务所为自己设计的办公空间是对杭州市中心一栋大楼旧有顶层空间的改造更新。原有顶层空间是大楼屋顶加建的仓库，包括电梯机房、消防楼梯间和倾斜的排水屋面。设计理念通过把握建筑材料的同质性、异质性以及时空性，创造旧有仓库空间和新办公空间的并置。

为了彰显正统建筑语言与建筑"方言"之间互动并行与辩证关系，设计保留、改造了一系列能够揭示其原有搭建的"非职业"的设计元素：钢构架屋顶通过白色喷漆处理，使原有锈蚀的混凝土梁柱依然作为结构核心被完全保留；房间采用架空的白色地胶板，5个镂空设计的地下橱窗玻璃镶嵌其中，展现了原有粗糙倾斜的排水屋面；从屋顶到仓库，从废弃到重新利用，疏散门的雨棚和粗糙的马赛克墙暗示着这片空间定义的变迁。

在这个延续、重组、再现的空间设计中，马赛克和裂缝质感的材料层次，与白墙的崭新交织出一种暧昧的凝视。褪色而带有多次修补痕迹的旧水泥墙体与白色新墙体的并置创造了另一个层次的异质性。

重工业感的电梯机房通过窗口展示成为事务所空间的一部分，延伸了时空感并创造了其自身的语境。坏损的橙红色疏散门扇被移植作为事务所会议室的大门，使旧建筑的元素得到尊重，焕发它旺盛的生命力。

零壹城市建筑事务所的空间设计，试图在与空间功能吻合的职业建筑设计和粗犷的原有非职业设计之间，创造一种辩证的对立关系。超越单一的设计风格，讴歌创作的不拘一格，这也是"零壹城市"的设计理念之一。

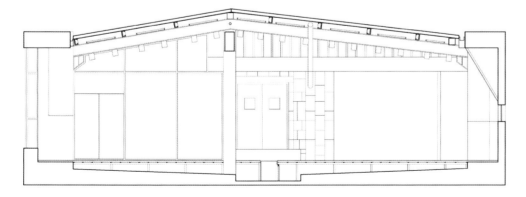

The new office of LYCS Architecture is a renovation design of a tower penthouse in downtown Hangzhou, with a panorama view of the West Lake World Heritage.

The original space with is a non-professional add-on storage room to the tower roof composed of elevator mechanical room, fire escape doors and slanted drainage surfaces. The idea of this design is to create the juxtaposition of this abandoned storage space and the new office, through the material homogeneity to heterogeneity and temporal attenuation of space.

In celebrating the collectiveness and parallelism of the authentic architecture language and the self-built "dialects", the design remains the imprint of several original elements in revealing its original non-professional property. The eroded columns were kept in contrast to the white steel frame of the roof. The elevated floor from the original slanted drainage roof is perforated and covered with removable glasses, in making the original roof surface explicit yet in depth. The fire escape door canopies and the rough ceramic wall are kept to imply the mutation from the abandoned storage space into an introverted architecture office, further to amplify the existing transformation from an exterior roof into an interior space.

The steel frame and walls are painted in creating a museum-like white space, as an exhibitive gesture to the remainders. Unlike a pure delicate white room, the painted walls with the original layer of mosaic and cracked texture give a level of ambiguity upon the completely new walls. The juxtaposition of fading ceramic wall segments with blocking and opening doors and the new white walls creates another level of heterogeneity.

The overlaying program of the heavy-industrial elevator mechanical room and the architecture office space attenuates time in creating its narratives. The broken orange roof door retired from its previous function as a guardian. It is moved thus respected as the interior door of the meeting room, in prolonging the lasting period of architectural elements and its space.

The design attempts to create the dialectical effect of architecture professionalism as the program is, and the non-professionalism in architecture. It celebrates multi-signatures over single signature, which also reveals one of the design philosophies of LYCS Architecture.

LYCS Architecture Office

华东 REGION EASTERN CHINA | 223

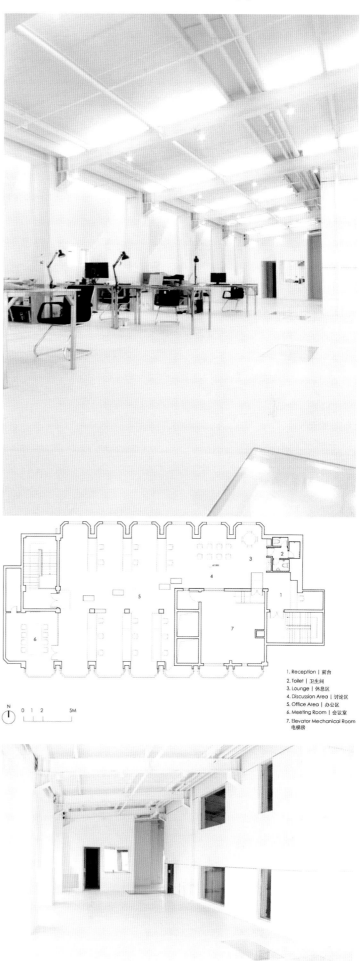

1. Reception | 前台
2. Toilet | 卫生间
3. Lounge | 休息区
4. Discussion Area | 讨论区
5. Office Area | 办公区
6. Meeting Room | 会议室
7. Elevator Mechanical Room 电梯房

杭州环球MUSE酒吧

项目地点：杭州
设计单位：ANS International Design & Consulting Pty Ltd
设计师：徐岭啸、汤昊隽、顾原涌、孙桢
建筑面积：1400 m²

项目坐落于杭州CBD中心，西湖文化广场环球中心四楼，ANS在这1500 m²的空间内打造了一个时尚前沿的高端俱乐部。

在这个罕有的10 m层高且无柱的空间内，设计从抽象的概念出发：若把同一平面分成两半，并弯折其中的一个平面，便可以形成一个三维空间。但在此，这种不可见的张力是通过每个面上不同材质的分割，线条的不同走向来实现的，包括彩色环氧地面，不锈钢墙面板及压塑铝板顶面。三维屏幕环绕中的DJ台，坐落于整个空间的最中心，从每一个角度都可以清楚的看到，也成为整个空间中的一大特色。卫生间的设计，运用了大量色彩鲜艳的嘴唇特写图案来制造出性感的氛围。

设计把这个"方盒子"里的各种功能最大化。整个"方盒子"被分成三个层面，最底下一层主要被设计成办公室，员工休息区，寄包区，厨房，卫生间和储物间等辅助空间。从一个由大理石铺设成的楼梯可以走到酒吧的二楼公共开放空间。这个空间设有大量灵活的卡座，考虑到酒吧的定位是针对一些高端客户，顶楼处设有三间VIP私人包房。调色玻璃墙面的运用不但可以保证空间的私密性，并可以在需要的时候，通过调节把玻璃从不透明变为透明，来增加与楼下空间的互动。

设计中最大的挑战来自如何设计加建层的钢结构，使得立体空间效果最大化。经过努力，我们欣慰的看到整个开敞空间的效果没有被任何结构柱所阻碍，VIP包房也好似悬浮在空中。

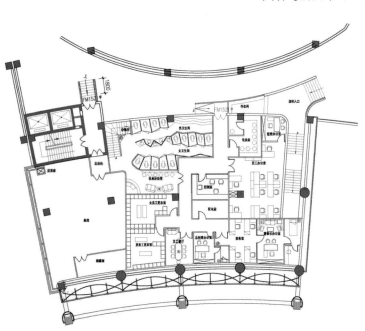

MUSE Global Hangzhou

Located on the 4th floor of the prestigious West Lake cultural plaza, ANS Interiors was commissioned to convert a 1,500-square meter space into a trendy, high-end dance club within the CBD of Hangzhou.

Given the rare nature of the 10m-tall, column-free space, the design concept originates from the abstract idea of pulling 2 identical planes apart while slightly twisting each plane, thus creating a three-dimensional cubical space. This invisible tension is imitated through the texture treatment on the various surfaces through the use of colored epoxy flooring, laminated wall panels and aluminum extrusions on the ceiling. The main feature of the space, the DJ booth, combined with a three-dimensional screen elevated and centrally located in the main open area, visible from all parts of the venue. The bathroom is designed with a bold combination of "lips" graphics to implicate a sexy yet contradicting perception.

The design begins with maximizing the area for various functions within the "cube."

In order to accommodate all of the back of house facilities such as offices, staff rest areas, lockers, kitchen, bathrooms and storage areas on its lowest level, the "cube" is divided into 3 different levels. The main open area of the club is located on the second level, accessed through an understated and marble-covered staircase. Most of the area of this level is occupied by generous and flexible booth seating, as the venue is targeted to high-end customers. 3 VIP private rooms and balcony booths are located on the top level. UMU glass panels provides VIPs the privacy and also, when desired, visual connections with the actions below upon switching the transparency from opaque to clear.

A lot of effort was spent on the structural steel frame design of the additional levels to maximize spatial quality and booth seating. As a result, the main open club area is unobstructed by any structural columns. The VIP rooms above seem to be hovering in mid-air.

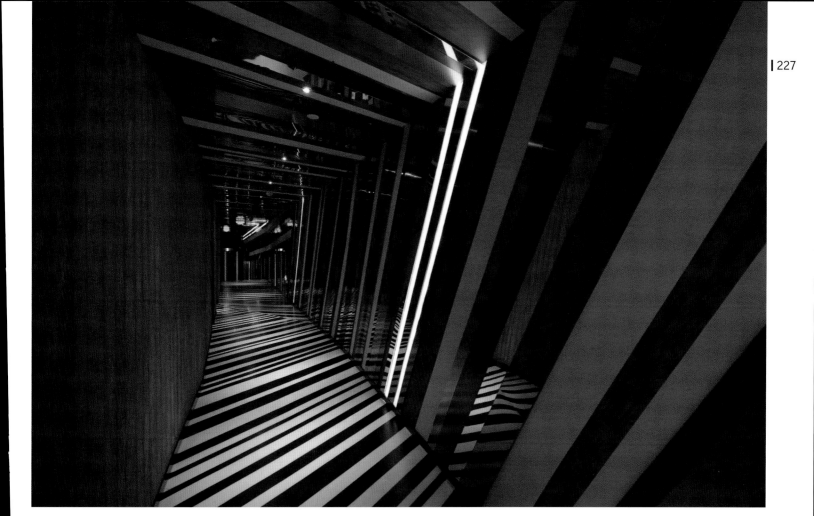

嘉兴月河客栈

项目地点：嘉兴市
设计单位：广州集美组室内设计工程有限公司
设计师：陈向京、曾莹
参与设计：皇甫丹琳、肖正恒
建筑面积：23000 m²
主要材料：金丝柚、黑胡桃木、伯利黄、山西黑
摄影师：罗文翰

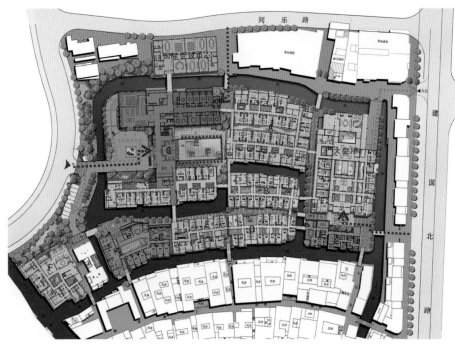

三面临水，八面来风。嘉兴作为江南文化的发源地，自古为富庶繁华之地。早在六千余年以前，先民便在此繁衍生息，孕育出璀璨的马家浜文化。遂历经"滨海泽国"、"嘉禾飘香"、"繁庶市镇"、"文风鼎盛"等重要历史时期。"滨海泽国"——水乡泽国，河网密布，是嘉兴自然环境的最贴切写照。"嘉禾飘香"——沃土嘉禾，为漕粮供应之地，稻作文化兴盛。"繁庶市镇"——明清时期，纺织业发达，市镇星罗棋布，实为丝绸之府。"文风鼎盛"——文人雅仕问鼎一时，风流人物传盛一世。经时代传承，这些历史时期的民风意象部分已升华为当地的民俗精粹，演绎成一抹别样的江南风情，成为嘉兴最吸引人之处。

在月河客栈的设计中，我们通过对当地建构手法的研究与继承，传统装饰纹样的抽象简化以及本土化材料的解析运用，有机的将禾香、庶市、文风、泽国，这些极具民风意向的主题融入室内空间，揉古释今，化凡为雅，营造出极具江南水乡风情并富有现代气息的主题体验空间。

Jiaxing Yuhe Inn

Surrounding by waters in three directiong, as the motherland of Jiangnan culture, Jiaxing is a flourishing city since ancient times. Six thousand years ago, people began live here and generated the famous Majiabang Culture. Then it has experienced "Coastal kindom", "Jiahe fragrance", "Flourishing town", "Great Culture" and other important historical period. "Coastal kindom"--waterside place, with internet rivers, is the most closly description of Jiaxing. "Jiahe fragrance"--fortile land of Jiahe, is a grain place, particularly of rice. "Flourishing town"- in Ming and Qing Dynasty, the spinning industry is flurish, fulfilled the town, which can be called as Home of Silk. "Great Culture"--here was full of talents and literati. After heritage from generation to generation, these customs and culture of history have been improved to the essences, performing a unique Jiangnan scenery and becoming the most attractive place of Jiaxing.

In the design of Yuhe Inn, through the study and inheritance of the local construction, simplification of the traditional decorative patterns as well as the utilization of the local materials, we have merged plant seeds, ancient market, cultural, water into this space. Combined ancient and modern together, converted general to elegance, we have created a most Jiangnan mood scenery with modern atmosphere.

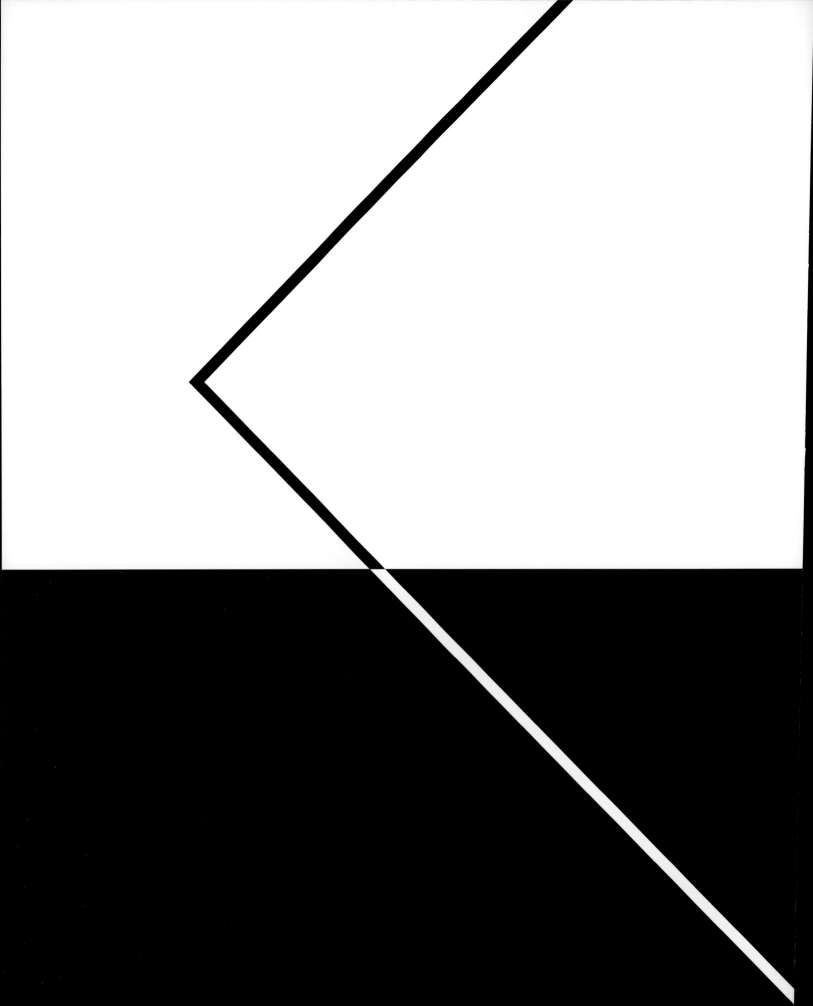

CITY
HONGKONG

香港

嘉禾黄埔电影院

项目地点: 香港红磡黄埔广场
设计单位: DPWT Design Ltd.
设计师: 陈轩明
项目时间: 2011
建筑面积: 5000 m²
主要材料: 喷漆玻璃、人造石、石膏板
摄影师: 陈志威

嘉禾黄埔电影院是针对普通顾客设计的电影院,公共区域的设计以简洁为主,大量运用了钢化玻璃和反光灯槽,增加了许多的亲和力。售票处是白色的柜台,底下有一圈发光灯带,整个售票处好似浮在半空中,如梦如幻。蓝色喷漆玻璃的背景墙,在灯光的映射下,清新自然。走廊处全部布置了发光的海报墙,比较醒目。等候区放置了许多不同色彩的小沙发,好似在白色的海洋里撒上了许多彩色的星星,让等候在此的顾客,心情得到了放松。卫生间的设计也别具一格。整个的设计,白色是主题色,配以彩色的家具和配饰,简约而不失活泼,重视细节的处理。

嘉禾黄埔电影院坐落于居民聚居区,毗邻大型购物广场,电影院的设计简单但不单调,满足了普通市民休闲娱乐的需要。

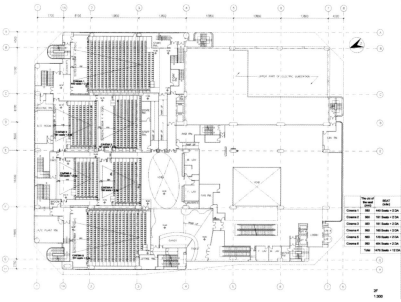

GH whampoa Cinema

The design of Golden Harvest Cinema at Whampoa is decorated in a style of minimalism. Large quantities of white and blue spray paint glasses are installed in the customer lobby. White marble tile are laid at floor so that It can blends together with the wall. On one side of the wall leading to the internal atrium, 4 TV showing the latest information of the box office are installed at the ceiling. They echo well with the TV installed above the box office counter. In response to the extensive broadcast of trailer, totally 8 numbers of 65 inches LED plasmas are installed on the wall to show the update movies and posters. This gives a very cinematic ambiance to the lobby. A candy bar is located in the internal atrium which provide international various delicious cuisine to the cinema goers. Colorful stools are scattered at the lobby to provide leisure and relaxing atmosphere.

Different palette hues are manifested in each cinema hall. Vertical fabrics with similar hue color are installed to highlight the effect and this can achieve a better acoustic effect. The blue carpet with colorful and star pattern blends of the different color as a whole.

The toilets are designed in a chic and modern style. Various ellipse patterns are painted on the ceiling to deliver a highlight touch to the design. The white color and extensive use of mirror in toilet provide a much larger visual effect.

皇室戏院

项目地点：香港铜锣湾皇室堡
建筑面积：**2200 m²**
设计单位：创智建筑师有限公司
　　　　　（AGC Design Ltd）
设计师：曾伟贤（Michael Tsang）
摄影师：邓家辉、林强健（Marcel Lam）

皇室戏院，香港第一间具有国际风范的完美影院设计，提供了3间影院，共250个席位，伫落在熙来攘往的铜锣湾中枢地区，以"跃层"的空间美学，演释着舒适、时尚、凝聚、亲和的空间剧本。

生活模式的改变，建立了另类的消费模式雏形，都市人日益减少在家及在工作上的时间，强化在外逗留的行为意识，酝酿出"家"的发酵环境与机会，建立逸品的个性标志，弥漫着感动生活的微熏，拥有一种沉淀在繁华、杂乱都市的凝聚魅力。

随着观众对品味的渴求与期盼，相对情景设定的环境演化将成为鼓动人心的流行创作，更多以人为本的环境制约服务标准，相继在空间中具体成型。皇室戏院是香港第一家逸品影院，凝聚了吟酿般的精华，体验从"聚"的集体氛围到"赏"的个人舒适感，一种与家融合的空间感受，近在咫尺的"弦外之居"。

在影院的设计手法中，注入了逸品的空间元素，如家中亲切、温馨的幸福调味，洋溢在一个舒适的环境中。"影院是家"是一种崭新、亲切的美学诠释，解决当下的生活方式及意识形态，创造一个独特的空间飨宴。大堂是他们的起居室，流动的雕塑如攀藤般延伸至双层高的起居空间，以独特的感动与个性化标志，开启"花"与外界的对话，绽放出灿烂的词汇。而对外的露台，静静地凝视着街道，隔离着一切的烦嚣，纳出一个愉悦的休闲、聚集场所。影厅是他们的卧室，宽敞的空间配置，先进的视觉效果及音响系统，豪华得有如身处于家里的沙发上，享受着银幕所带来的震撼与感动。

一个玻璃纤维复合材料的组合，从腼腆、朴实的大理石墙核心不断增长，衍生微妙的开花现象。表面的光线就像个热烈的舞者，在不同角度和不同的媒介中，交互起舞，催化三维轮廓的视觉效果，更具吸引力。花的绽放，诉说着无尽的话语，触碰着，渗透着，每一个空间及每一个紧密环扣的细节，弹指之间散发着自然、清纯的"绿色"气息。在进入影院之际，采用发光膜材料在天花延伸，扭动，拉扯出诱人的身躯，流动着空间的韵味。进而，以风化锌板编织出流畅的纹理，纵横交错于花的形体之中，透露出逸品优雅的空间触感。而毗邻在旁的玻璃垂幕，如珍珠点点般耀眼夺目，就像为盛开的花朵提供了一个完美的肩膀，在闪烁倒影中，静静地倚靠着涟漪般的细腻。

Boutique Cinema

This is the first boutique cinema in Hong Kong, offering 'loft-design' aesthetic with comfort, stylish and homely in a central location at the bustling and hustling Causeway Bay area. The cinema comprises 3 Houses of total 250 seats.

This "Boutique" design concept is a strategic solution addressing the emerging consumer behavior pattern of urban people spending less and less time at home, not necessarily having longer working hour, but more willing to spend time with friends outside home. The application of "Boutique-quality" is more often an icon of personality touched style and is a means of living to stay humanely in the bustling & hustling city."Boutique" as Home As increasing demand on tasteful cinema audience, such particular emotional environment or "boutique feel" design will become trendy for creating more humane, creature comforts & personal cinema's service. Windsor is the first boutique cinema in Hong Kong with all essences of boutique experience- "feel like at home", from gathering patio area to high-end comfy sitting, it is all about to provide & feel like home away from home."Home" away from home Cinema is their Home. This aesthetic homely atmosphere illustrates a new meaning, addressing the needs of current lifestyle and the urban people, creating a unique sensory experience.

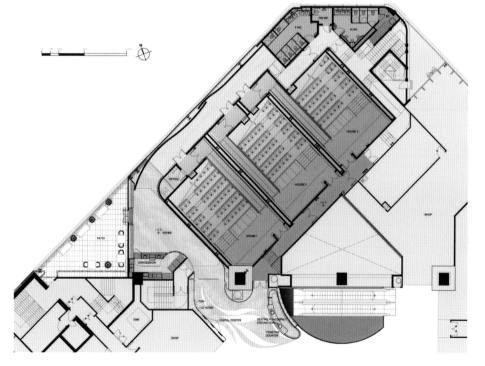

"Foyer as Living Room", a blossoming flower from wall to ceiling creates dialogue with the outside world, while separated from the hustling & bustling city; "Auditoria as bedrooms", where spacious couch-like seating plus state of the art AV systems, creating a trendy home-like cinematic experience for those in search of a home away from home.

A subtle blossoming phenomenon, made out of GRP composite, inherently growth from earthy marble wall is the centre piece of the design. Light reflecting off the surfaces at various angle and different planes are tuned to optimize the visual effect of making the 3D profile of more reveal. The blossoming effect penetrates into every stage of boutique experience from the box office to every design details in the cinema premises so that it is overwhelming with this dynamic "Green" live. The flowing per-weathered zinc sheet ribbon crisscrossed between the sculptural flowers creates an accent of boutique senses. In contrast with the glass-feature screen,provides a perfect backdrop for the blossoming flower lay next to the twinkling water reflection.

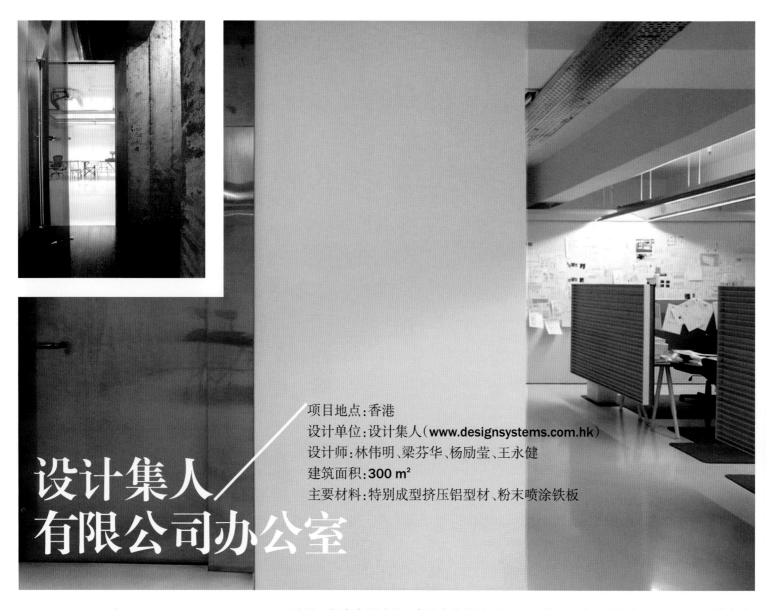

设计集人
有限公司办公室

项目地点：香港
设计单位：设计集人（www.designsystems.com.hk）
设计师：林伟明、梁芬华、杨励莹、王永健
建筑面积：**300 m²**
主要材料：特别成型挤压铝型材、粉末喷涂铁板

" life is a leaf of paper white, thereon each of us may write his word or two."
　　　　　　　　　　　　　　　　-- Amy Lowell

　　这是一间室内设计公司办公室的设计项目。"白纸"是设计的主题，比喻公司作为一个白色的世界等待着设计师为它增加色彩。在设计的范畴里，每一个项目开始时也像是一张白纸，随着项目的发展，白墙和白板上都开始渐渐地贴满了研究计划、草图和项目表，记录了每个项目的演变，就如一本设计师的日记本。铜制的大门和白色的地板收集了每个员工和客人的指纹和脚印，亦成为了公司发展的见证。

　　为了迎合设计事务所所需的大量储存空间，在墙身白板趟门后面隐藏了很多储存架，把设计师们杂乱的图纸遮盖起来，这样的设计保证了表面的整洁和条理性。虽然设计选用的物料价钱都比较便宜，但是都经过特别处理。例如，将众多杂乱的电线用传统玩具"扭蛋"的塑料外壳包裹着，特别的扭蛋造型也成为办公室里的装饰品；还有，出于环保考虑，所有办公桌都选用可循环的木纤维甲板为材料。

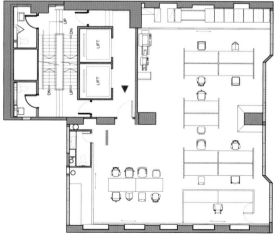

The Office of Design Systems Ltd

This project is the office of an interior design firm. Central to the design concept are originality, spatiality, and functionality.

Every design begins with a blank sheet, ready for colours brought by every design decision. In a stripped clean lobby stands a brass door behind which a 300-square-metre office full of crispy white daylight awaits. Inside, the ceilings, walls, pin boards and floors are all in white just to constitute a white world for designers.

The canvas-like interior materializes the design concept and creates opportunities for myriad exhibition of design process. As each project progresses, the walls and boards are covered with researches, design sketches and presentations, recording the evolution and development of every design stage. Eventually, such accumulation of works can be compiled as the designers' diary. Similarly, the white floors and brass door, a unique and daring attempt for robust daily use, collect the footprints and fingerprints from the staff and clients, witnessing the growth of the company.

Due to the chaotic routine of a design firm, the designers see the need for a new concept of working disposition. The realization of the project delivers a clever combination of the ideal solutions to two very contradicting needs – maximum hidden surfaces by "the mess concealer" and maximum open surface by "the mess promoter". Thus, the installation of plentiful storage is essential for such elasticity of display. Mess is readily covered up behind the sliding boards. In turn, these white covers exhibit the effort being spent at any given moment. The office will become a clean plate after each project, ready for a new one to commence.

项目地点：香港东岛城市大厦4号地铺
设计单位：Joey Ho 设计有限公司
设计师：Joey Ho
参与设计：Alfred Kwok、Fai Wong
建筑面积：122 m²
主要材料：油漆、地毯、织物、皮革、磨砂玻璃、普通玻璃、不锈钢、石子、墙纸、实木地板、复合木地板
摄影师：Graham Uden、Ray Lau

Azura是香港太古地产有限公司打造的豪宅，它座落于西半山，是香港知名高端区，这里通往中央商务区及附近文化娱乐景点的交通十分便捷。

该住宅区的名字Azura（汉语意思"蔚蓝"）是晴朗天空的颜色，它自然使人联想起大自然的美丽和简约。这个名字触发了设计师们要为人们创造一种简约生活的灵感，从而远离城市生活的繁忙和喧哗。设计师们采用了明快的设计主基调，他们相信纯净简单可以成为一种奢华的形式。他们旨在创造一种氛围，供看房者重新审视生活，在欣赏样板间的同时，卸去繁忙生活带给人们的各种沉重压力。

一进入样板间，看房者首先将通过一个幽静的小径，沿着弧形的墙面可以欣赏到不断变换的自然风景。这是一片规划中的海面，海面上不时有海鸥飞过，勾起来访者对大自然的遐想。整个小区的氛围宁静浪漫，使人仿佛置身于夜色之中，它意味着忙碌的白天结束，世外桃源般夜晚来临。

样板间自然的泥土色调给人一种安静优雅、不张扬的美感。客厅独特的落地窗旨在给主人一种空间感，也有利于自然光和新鲜空气的进入，另外，通过这些落地窗也可以欣赏香港如画般的美景。时尚的半开放式厨房（配吧台）使主人在工作了一天之后能有一个舒适的进餐用膳的环境。现代化卫生间突出其开放感，省去了不必要的约束，从而使主人可以身心放松地使用卫生间。建筑师们在Azura样板间设计上展示了最现代的简约派居住环境。

经过令人神清气爽的小径，映入眼帘的是内部花园，在柔和光线的掩映下，可以看到花园中的竹和各种植物，这里是来访者进入另一个空间的过渡。销售部的设计也体现大自然的感觉，它坐落在一片花园之中，四周长满绿色植物，而地毯上则点缀着设想中的图案。

通过展示世外桃源般的宁静、祥和、舒适，设计师们希望来访者能够产生真正拥有他们未来家的憧憬，并希望样板间能帮助他们找回大自然的平衡。

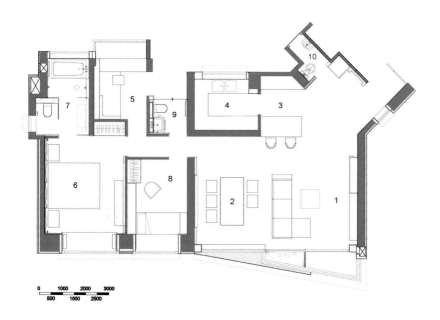

Azura

Azura is a luxury residence from Swire Properties Limited. It situates in Mid-Levels, a well-established and high-end residential district on Hong Kong Island with easy access to the commercial business district as well as local culture and entertainment scene nearby.

Naming as Azura ("sky blue" in Chinese) which means the color of serenity, it lexically connects to the beauty of nature and a simpler order of things. This triggers the inspiration for the designers to create a simple living for people to overcome the hustle and bustle in the city. Using a clean-cut approach as the key design element, they believe pureness and simplicity can be a form of luxury. They aim to create an atmosphere for visitors to review life itself and relieve all sorts of burden during the entire spiritual journey of the show flat.

At the beginning of the show flat experience, visitors will walk through a quiet path with a projection of a moving landscape of the nature along the curved wall. It is a conceptual sea with birds flying periodically for fully arousing visitors' natural senses. The mood of the place is specially set to be serene and romantic, as if a night scene, which symbolises the end of a hectic working day and start of a mellow evening in a sanctuary.

Heading to the show flat, its natural earth tone presents quiet elegance and understated beauty. The unique architectural floor-to-ceiling windows in the living room are designed for space, natural lights and fresh breezes. They, at the same time, portray the picturesque view of Hong Kong. The stylish semi-open kitchen with a bar table offers a cosy area for food and drinks after a long day in work. The modern bathroom has the emphasis of openness for lessening the sense of constraint. By this, it enables a restful bath to perfectly unwind oneself. The designers demonstrate a state-of-the-art minimalist living environment in this show flat of Azura.

After the refreshing journey, an interior gar-

den is constructed with bamboo-like rods and plants under lustrous light as a transaction before visitors reaching another unit. Closely connecting to the nature concept, the sales office is built in a garden setting with green plants surrounding and carpets with conceptual pattern of landscape.

By presenting a sanctuary of peace, calm and comfort, the designers hope that visitors can be enlightened for the genuine desire of their future home and that the show flat can restore their natural balance after the journey.

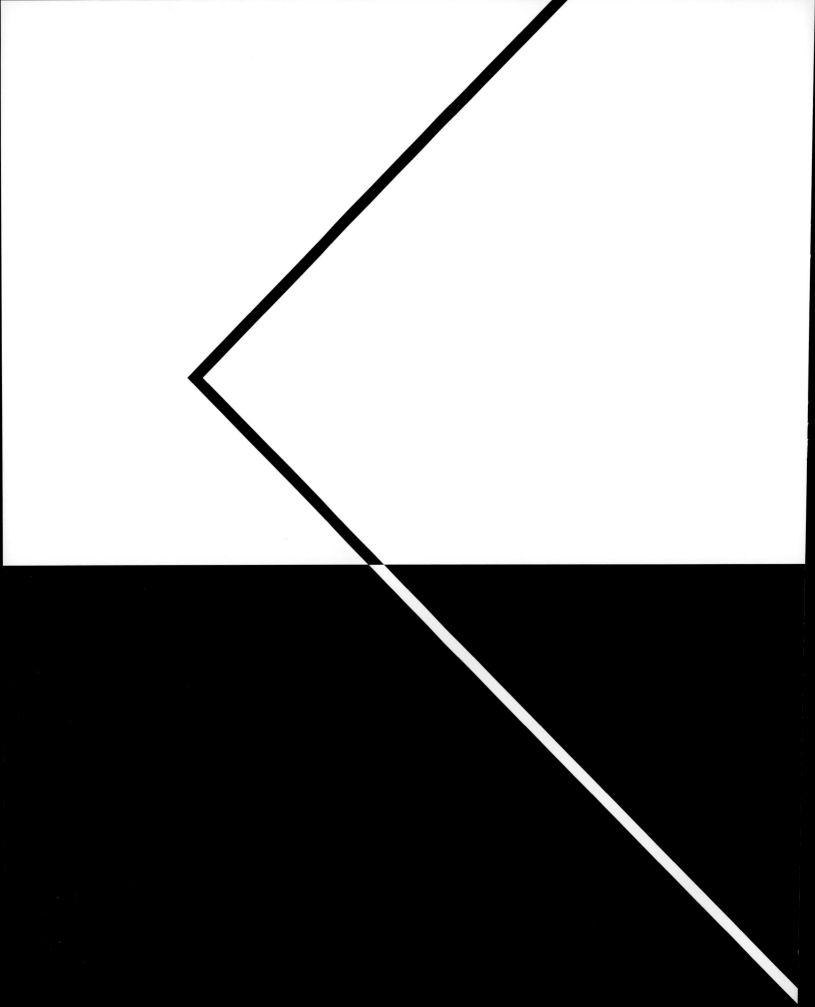

REGION 台湾
TAIWAN

台北W饭店

项目地点：台北
设计单位：G.A. Design International Ltd.

台北W饭店坐落于素有"台北曼哈顿"之称的商业旺地——信义区，可观七星山和阳明山之秀丽美景，可享大台北动感十足之都市生活。

台北W饭店楼高31层，由伦敦著名建筑事务所G.A. Design International Ltd.倾力打造，周身采用玻璃外观，是城内唯一能纵览信义区繁华街道全景的酒店，亦可一睹以逾500 m身量直耸云霄的地标建筑、全球最高楼之一101大厦之芳容。

来到台北W饭店，宾客们定将对这个由坚固不锈钢及镜面设计构成的The Chain(锁链)惊叹不已，它象征了这间酒店对台北的热情与厚爱。此外，一面巨大的绿色墙壁上种满了来自台湾的各种植被，令人耳目一新。步入其中，宾客将开启一段体验台北W饭店独特魅力的全感官之旅。台北W饭店彰显了多种领先的设计布局，将感官享受与空间设计完美结合起来。

互动灯光设施可与走进酒店的宾客互动，随着环境的变化呈现不同的形状、外观及感觉。出自W酒店未来设计师rAndom International之手的幻光魅影（To Light You Fade）位于台北W饭店的地下层，以一面再生木头墙为背景，宾客可与灯光肆意共舞。该设施配有特制软件，由数百支独特的OLED（有机发光二极管）组成，它们个个出身名门，均来自德国亚琛的全球首条生产线。头顶上方是一个由塑料与竹简构成的精美雕塑，倒映在镜面之中，施施然由大厅天花板悬垂下来；而"Purple Target"则是

一幅镶满优质图钉的帆布,迎接着来到信义中心地带的宾客。安装在入口处及十楼地板上的白色投射灯给人以水雾迷蒙的感觉,走过这里便是独具一格的W起居室,可以欣赏到WET泳池摩天台尽头的标志性泡沫雕塑。

走进W Living Room,在实木板材堆栈而成的迎宾台前,铺有一幅用红色电子花束点缀的地毯。Living Room摆放着宽敞的圆形睡椅,一扇扇实木百叶窗组成了错落有致的檐篷,若将檐篷折低,便合围而成二楼的会议区。Living Room侧面装有一个11 m高的"W"型屏幕,对面是一个同等高度的壁炉,为您在台北的寒冷夜晚提供一个时尚温馨的环境。宾客还可在由木条拼接的高大茧状座椅营造的私密空间中放松自己,6 m长的大理石吧台将为您随时提供各式果汁及饮料。W Living Room内里有一整面墙完全由玻璃构成,形成内外通透视野,穿过光滑的玻璃嵌板便可直达WET泳池摩天台。

台北W饭店的特色WOOBAR(与酒店的迎宾前厅并列)的地板以奢华的天然突尼斯石材铺就,取名为"Autumn Brown"("秋色棕影"),造型独特的圆形白色真皮宽大软凳一字排开,与入口地灯的水滴设计及泳池的水花飞溅遥相呼应。WOOBAR层高10 m,顶棚可随时间进行调整,大厅一端是一座高科技壁炉,另一端则是DJ台,世界顶级DJ将透过操控这些与设计浑然一体的先进音响及灯光系统,给眼光独到的泡吧一族最畅快的享受。WET泳池摩天台清洁明亮,四周环以木栈道,设有开

放式壁炉,绿植葱郁,更修建了生态绿墙。泳池一端是一座醒目的金属泡沫雕塑,与银色水珠交相辉映,一群形态各异的陶制蝴蝶振翅欲飞。

10楼的 the kitchen table 餐厅采用明亮轻快的花园别墅设计,并添加一丝现代气息,内部装修以鲜明的嫩黄色与清爽的白色为主,犹如清晨的阳光洒满房间。the kitchen table 在设计之时考虑到了台北的亚热带气候,两面墙壁均安装了超高滑动门,直通毗邻的 WET 泳池摩天台露天就餐区。奇幻的雕塑顶棚就像一个藤蔓盘绕的篱架,装饰简约的混凝土柜台又仿佛一张张公园长凳。互动的开放式厨房中央是一个 2.7 m 长定制 Moletini 炉灶。The kitchen table 的其它墙面均以书架覆盖,上面摆放着装满阳光的罐子——每一个罐子都是一支太阳能 LED 灯泡,白天可搬到 WET 泳池摩天台上充电,晚上再运回餐厅在架子上发光。

位于台北 W 饭店的最高层 31 层的紫艳中餐厅,提供新广东料理,亦是 W 饭店的全球首间中式餐厅。紫艳中餐厅的落地大窗将信义区的繁华街景一览无遗,夜色中的霓虹灯流一直蔓延至台北周遭的远处山脉。紫艳中餐厅展示取材自中国传统烹饪工具的原创艺术品,创意大胆前卫。有一个大型墙面装置是由数百个金属汤勺组成,排成立体圆圈;另一个引人瞩目的装置则是由中式瓷制汤匙组成,造型错综复杂。数千个形状各异、大小不一的饼干模具以独特造型组成两个并列装置,而餐厅侧面则是一组由创意十足的中式厨具组成的墙面雕塑。

紫艳中餐厅墙面以深紫色亮漆与光亮的望加西黑檀结合,隐匿在夜色之中,烘托出无限美景。厨房中央是一个叠层厨房,外镶黑色大理石,每当夜幕降临,烹饪区便忙碌起来。座位区摆放着宽大梳,排列为云层状,周围环绕由粗绳织就的高达 5 m 的屏风。匠心独运的屏风,恰似中国的鸟笼,与之遥相呼应的是不锈钢餐台上高挂的丝线灯笼。酒吧偏居于 31 层的一隅,由玻璃幕墙围绕而成,堪称台北最

气派的场所之一。酒吧外墙由一面面多棱镜组成,在双层高的紫水晶屋中闪闪发光,给食客带来新鲜的用餐体验。一套专为台北W饭店打造的"飞碟灯",布满酒吧的屋顶,透过温馨的灯光,宏伟的台北101映入眼帘。

台北W饭店拥有405间客房和套房,使其跻身台北规模最大的奢华饭店之列,每一处私人空间都可欣赏无以伦比的都市美景。暖色调石料、抛光实木、绿意盎然的植物雕花地毯与中式灯笼罩子中充满现代感的精致灯光形成了令人沉醉的奇妙对比。所有客房均设有室内阳台/休闲区,宾客可以在台北这座国际大都会中享受一份属于自己的宁静,350纱支密度亚麻的W特色睡床将为您带来一夜酣眠,优雅现代的白色书桌搭配有符合人体工学的皮椅。浴室配有超大海岛风情浴缸,与以地铁为灵感的红色或黄绿色壁面砖及木质隔断形成对比,让您放松身心。考虑到国际飞行旅客的需求,每间客房均配备了高科技设施,包括:高速有线及无线互联网接入;42英寸平板液晶电视;Bose声道系统;iPod充电座;提供语音邮件服务的IP电话以及为客房增添更多奇妙色彩的W特色十二生肖。不少客房进行了台北式的改造,木墙上悬挂着题为"你在哪里"("Where Are You")的解构地图,窗外一边是阳明山的自然美景,一边是台北的万家灯火。解构地图正是电力十足的台北的缩影,给宾客带来一丝逃脱喧闹的幽静之感。

Taipei W Restaurant

Taipei W Restaurant is located at the prosperous business area known as "Taipei Manhattan"—Xinyi District. You can enjoy the beautiful landscape of Qixing Mountain and Mingyang Mountain National Park as well as the dynamic urban life of Taipei.

Taipei W Restaurant is 31 floors high, which is designed by famous architecture institute of G. A. Design International Ltd. in London. Its entire body is made from glass, which is the only hotel that can enjoy the whole prosperous landscape of Xinyi District. It also allows you to view the 500 meters high landmark, the face of 101 Building which is one of the highest building in the world.

Entering Taipei W Restaurant, visitors will surely surprise for the Chain consisting of stainless steel and mirror design. It symbolizes the passion and love for Taipei of this hotel. Furthermore, there are "planting" various of Taiwan plants on a huge green wall, which is much fresh to people. Enter it, visitor will have a great senses experience of Taipei W Restaurant's special charming.

The interactive light device coupled with the walking people, shows different type, appearance and feeling with the changing environment. To Light You Fade created by rAndom Internationa of W Restaurant is located at the underground. Coupled with a recycled wooden background, visitors can dance with light easily. This device is equipped with special soft ware. It is consisted with hundreds of unique OLED (organic light emitting diode). They are all well-produced from the world's

first production line of Aachen, Germany. On the head, there is an exquisite sculpture of plastic and bamboo. Reflecting on the mirror, they look like hanging from the ceiling. Furthermore, Purple Targe is a canvas studded with high-quality pushipin to meet the guests of Xinyi Center. The white spotlight installed at the entrance and the 10th floor give us a dreamy sense. Passing through it, there is a unique and special W Living Room, where you can enjoy the WET swimming pool platform's bubble sculpture.

Entering W Living Room, you will see the wooden reception table and the carpet paved with red electronic bouquet pattern. There is placed a huge round couch, wooden shutters in the Living Room. They are placed in orderly. If put the canopies lower, then they will form a conference area at the second floor. At the side part of the Living Room, there is installed a W type screen with 11 meters high. Facing it was a fireplace with same height, which will provide you a fashion and warm place in cold winter. Guests also can relax themselves in the huge cocoon-like seat made of wooden strips. You can enjoy the fruit juices and drinks at the marble bar with 6 meters high the time you like. There is one side of wall in W Living Room made of glass completely, forming a transparent vision. Passing through the glass panels, you will come to WET swimming pool directly.

Taipei W Restaurant's unique WOOBAR (in the same line with the reception table of hotel) is paved with the luxury natural Tunisi stone, named as "Autumn Brown". The special round white leather soft chair are placed in order, which is echoing with the drop design at the entrance and the spraying water in the swimming pool. WOOBAR is 10 meters high, which can adjust the roof by time. There is a high-tech fireplace at end of the hall. At the other end, there is a DJ table. The world-class DJ will be played here. Coupled with the advanced stereo and the light, you will have a great vision experience. WET swimming pool is clear and bright, surrounding with wooden road. There is set an open fireplace. Seeing the flourish trees, it is like entering a ecology park.

At one end of the swimming pool, there is an shining golden bubble sculpture, echoing with the silver water drip, looking like a group of butterflies.

The kitchen table at tenth floor is adopted bright and lovely garden villa design. Added with some modern art sense, the internal decoration is adopted the bright yellow and white as the theme, which seems like the morning sunshine. When designing the kitchen table, considering the subtropical climate of Taipei, the two walls are all installed with ultra-high sliding doors, straight to the WET swimming pool platform for food. The magical sculpture roof is like trellis vines. The simply concrete counter is like benched in park. At the interactive open kitchen central is a 2.7 meters long Moletini stove. The other walls of the kitchen table are covered by bookshelves. There are placed many sunshine jars—each jar is a solar LED light bulbs. During the daytime, they can be moved to the WET swimming platform for charging and then take back at night.

The Purple Chinese Restaurant located at 31the top floor of this hotel provides Guangdong cuisine, which is also the first Chinese restaurant of W. The large windows of Purple Chinese Restaurant have absorbed the bustling street of Xinyi District completely. The neon lights have spread to the distant mountains around Taipei. The exhibition materials of Purple Chinese Restaurant come from Chinese traditional cooking tools, with boldness creation. There is a large wall made of metal spoon and arranged in three-dimensional circle. The other eye-catching device is made of Chinese porcelain spoon. Thousands of various shapes and sizes biscuit molds are made into two parallel devices in special shape. The other side of the restaurant is a group of creative Chinese kitchen wall sculpture.

The wall of Purple Chinese Restaurant is made of purple lacquer and bright Garcia ebony, hidden ii the night, seeming much beautiful. The central of kitchen is a laminated kitchen, which is inlaid black marble outside. When night falls, the cooking area will be busy. The seats are placed in a spacious way. They are surrounded by five meters screen made of rope. This excellent screen likes a birdcage in China. Echoing with it is a silk lantern hanging above stainless steel dining table. The bar is located at a corner of the 31th floor, surrounding with glass walls, which can be called as the most impressive place. The outside wall of bar is made of many prism mirrors, shinning among the double height amethyst house. It will bring a fresh dining experience to visitors. A set "UFO" lights made for Taipei W Restaurant covers the whole roof of the bar, passing through the warm light, the grand Taipei 101 will come to your eyes.

The Taipei W Restaurant possesses 405 rooms and suits, making it be listed in the largest luxury restaurant in Taipei. Each private space allows you to enjoy the beauty of this urban. The warm color stone, polished wooden, green plant sculpture and the Chinese lantern cover are all full of modern art sense, letting people immerse into

such nice atmosphere. All the rooms are equipped with balcony/leisure area. The guests can enjoy a piece of quiet of themselves here. 350 pieces of thread-count linen W beds will bring you a refreshing night. The elegant modern white desk is coupled with manmade leather chairs. The bathroom is equipped with a large island-style bathtub. They have formed a contrast with the subway-inspired red or yellow-green wall as well as wood partition, allowing you have a fully relaxation. Considering of the needs of international flight passengers, each room is equipped with high-tech facilities, including: high-speed wired and wireless internet access, 42-inch flat screen LCD TV, Bose-channel system, IPod cradle, the IP phone with voice mail service as well as the wonderful funny Zodiac. Many rooms have been transformed in Taipei style. There is a deconstructed map named "Where Are You" hanging on the wall. Seeing through the window, there is a beautiful Yangming Mountain and houses of families. This deconstructed map is just the microcosm of Taipei, bringing a sense of peace to visitors.

大城朗云接待中心

项目地点：台湾台中
设计单位：暄品设计工程顾问公司
设计师：朱柏仰
参与设计：刘益志、范万遵
建筑面积：1173 m²
主要材料：木作、石材、铁件、茶镜
摄影师：李国民

在"大城朗云"的公共空间设计里，要完成一种再尝试：将现代艺术(Modern Art)之几何构成所呈现的感官要素融入三度空间；塑造一种新的空间隐喻(Metaphor)——现代都会的成熟品味。

以塑造"都市景观"为设计原点，开放式之市内建筑呼应都市街廊；以呈现现代艺术之线条、板块分割、构图构成整体三度空间为整体设计概念。

本案作为接待中心，企图以室内外互动之建筑体，具艺术性格之视觉，空间形貌来呈现存在于都市环境(Context)中的建筑体表面与室内融合为整体，互相渗透的空间模式。

本案"墙"以作为物件的概念：几何线条构图的透明玻璃台；玻璃与内庭、水池、坡道穿插融合的实体板墙；板块构图，结合室内主体厅天花集合构建、楼梯、中庭；整体以三度空间作为都市街廊的视觉主面。

Dacheng Langyun Reception Center

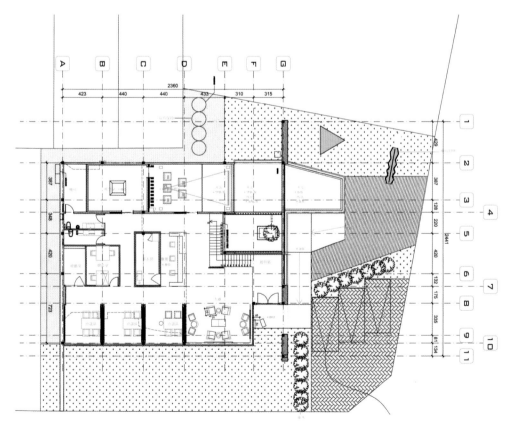

In the design of "Dacheng Langyun" public design, it need to finish another try, that is merging the Modern Art's geometrical expression into the three-dimensional space to shape a new space metaphor—modern city's mature taste.

Taking shaping "urban landscape" as the starting point of the design, the open style echoes with the urban street. The modern art lines, blocks and the composition are the whole design concept.

As the reception center, this case attempt to combine the indoor and outdoor function together. The art character is the visual aspect. The space appearance is used to show the urban context. The construction appearance is combined with the indoor part, penetrating with each other. In this case, "wall", as the concept of object, is a transparent glass unit with geometric lines. It is an real wall that separate the glass and outdoor. Then insert into the indoor, pond and slope. It is a kind of plate. Combined with the amin hall ceilings, stairs, atrium, the overall appearance is a kind of three-dimensional as urban street gallery.

草山水美

项目地点：台湾台北
设计单位：硕合室内装修股份有限公司
设计师：黄严仕、江姿莹、许珮填、蔡健原
建筑面积：200 m²
主要材料：缅甸柚木、铁件烤漆、磐多魔、珊瑚蓝、银灰洞石
摄影师：Sam + Yvonne／吴启民

一对兄弟，各自不定时需要一个安静的驿所停留……

从事精密外科手术的哥哥，频繁往返两岸拍戏的弟弟，工作的劳顿、旅途的风霜，都需要在这个工作之外的生活容器里，得到深度的休息以恢复耗损的精神与思绪。于是，借由动线的区划、活动隔屏的收放，达到空间的开放或隐蔽，这个生活容器，可顿时成为层次分明的一人住所、二人公寓、或一群好友聚集的派对空间……

空间的流动性成为贯穿、融合建筑室内与户外的主轴。在布局上，顺应窗外山景与采光条件，拉出垂直与水平的轴线与层次，置入长桌区及沙发区，利用入口玄关两侧隐藏式的滑门连结或关闭两处空间。长桌区旁的全高书柜，订制的抽屉大小刚好是剧本专用，不定时和导演编剧激发灵感、讨论剧情，大面的落地窗，将露台水池交叠北投大屯山风景框入室内。沙发区背墙是整面的DVD柜，让喜爱电影的兄弟二人可尽情收藏。朋友们来时，柚木实木拼贴的长廊露台串联起两端，成为完整的PARTY空间。

拥有北投天然温泉的汤区，利用土、木、水的结合，天然凿痕的石材表面，柚木屏风，从天而降的雨淋，充分沉淀心灵。经过与自然的对话，彻底放松。

空间是充满记忆的，它清楚地刻划居住者的使用习惯。整个空间透过物件细节的呈现，诉说材质本身的美与特性，并满足屋主精神上的需求与品味的彰显。

1. FOYER
2. TOILET
3. LIVING ROOM
4. DEN
5. HOT SPRING
6. BALCONY
7. POND
8. SHOWER ROOM
9. BEDROOM
10. WALK-IN CLOSET
11. DINING ROOM
12. KITCHEN

HOUSE FOR 2

A pair of brothers, they stay on a quiet place from time to time…

The old brother engaged in precision surgery and the young brother playing movies between Taiwan and Mainland China, hard working, tired journey, they both need to have a fully relaxation and restore their spirit and mind in daily. Therefore, upholding the moving lines separation, and the retractable partition, they have created an open or hidden space. Such life box can be divided into one-person home, two apartments for two people, or even a party space for group…

The mobility of the space becomes the axis of passing through the indoor and outdoor. In terms of the layout, they are conforming to the mountains and light conditions, pulling out vertical, horizontal axis levels. The designer has put long table area and sofa into it, using a porch at the entrance for connecting or closing. The high bookshelf beside the long table, the customized drawer is just suitable for book. Communicating with director to gain inspiration, discussing plot, and large floor window has absorbed the open pool and Beitou Datunshan Mountain landscape into space. The back ball of sofa area is the entire DVD cabinet, letting the movie lovers can enjoy the favorite movie collection. If friends come, the teak wood will be connected with the open corridor, forming a complete party space.

Possessing the natural hot spring in Beitou, combining the soil, wood and water, the natural chiseled stone surface, and teak screens, the rain from heaven, they all fully comfort our heart and soul. Through communicating with nature, we will get a fully relaxation.

The space is filled with memories. It records the habits of the occupants clearly. Through the detailed objects, the entire space is showing a beauty and character of the material itself and satisfies the needs of the house owner as well as his characteristics.

西园29服饰创作地

项目地点：台湾 台北
设计单位：建构线设计有限公司
设计师：沈志忠、李怡霖
建筑面积：一层 291.62 m²
　　　　　二层 271.96 m²
　　　　　三层 303.62 m²
主要材料：EROXY、铁件喷漆、防火皮革、
　　　　　人造石、清玻璃、PVC地板、
　　　　　柚木木皮染色

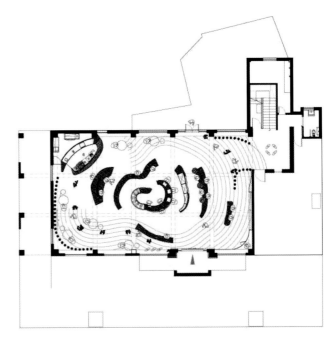

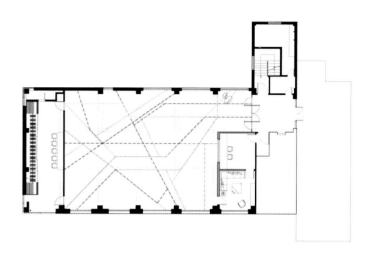

　　服装设计是以人的形体与肌肤为考量的，设计中所运用的材料也都蕴含着有机体，因此我们设定大自然的生命体——树，为此案的设计主轴。其中万华服饰文化会馆将化身为万华最重要的一棵树，扮演着培育台湾国际服装设计人才的角色。

　　空间的设计概念以一种有机的形态出现，并逐渐展开，象征服装设计的多元发展与其蓄势待发的创意能量。

　　进入一楼接待厅的空间犹如进到树的内部。该空间的设计概念取自于树体的结构，借由大树年轮的轮廓概念来组建和构成空间的形态，一圈又一圈的室内排布，如同大树的年轮一般，赋予了室内应有的空间机能。在垂直意象上，则采用了植物体的"维管束"概念，维管束具备养分与能量传输的功用，在此我们将维管束定义为发光体，将光的能量传送到空间中。

　　二楼的工作室区域，以"树中巢"的象征意涵起头，意指服装设计须有蜂巢式缜密的心思。我们在大树的空间中建起一格格的玻璃巢房。

　　巢房里孕育着结晶种子——服装设计新秀，设计师以此为出发点，找出文化馆空间里最适合的角度与元素，让它们贯穿于空间中的垂直或水平的方向，服饰文化馆的创造力与影响力自然贯穿于其中。

　　三楼为多功能活动展演空间，以万华地区街道的轮廓做为灵感源泉，将街道的形象转换为展板的挂轨形式反映到天花板上，可多功能运用、弹性十足的设计概念，让空间可以根据实际的需求进行多元的变化，充分发挥其视听与展示功能，成为这里最大使用效益的展演舞台。

　　在色彩的运用上，一层白色为主题，展现空间纯净的气质，朴素与华美并存，个性与包容并重。二层的工作空间用色沉稳大方，为员工营造最优质的空间氛围。

　　本案的设计概念新颖独特，三层的空间功能明确，分别以"树"、"蜂巢"和"万华地区街道"三个不同的设计概念展开，结合实际情况，进行设计意向和概念的表达。层层深入，精心设计，最终构建了一个实用与美观并存的时尚创意空间，令人叹服。

Fashion Institute Taipei

Costume design takes human body and skin as the basic. In design, the materials used are also containing the organism. Therefore, we set the life of nature—tree as the axis of this case. The Wanhua Costume Cultural Center will be embodied as an important tree, playing an role of developing Taiwan International Fashion Design Talent.

The design concept of the space is shown in an organism and extended gradually, which symbols the multi-elements development and creative energy in consume design.

Entering the reception hall at the first floor, you will feel like in a tree. This design concept comes from the tree's structure. Taking the tree ring shape as the concept to create and form this space, the rings are laid one by one, liking tree's ring, endowing a proper function to space. On vertical imagery, we adopt the "vascular plant" of plant. Vascular possesses the nutrient and energy transferring function. Here, we define the vascular

as the luminous body, delivering the light to the space.

At the office area of second floor, taking "room in tree" concept as the starting point. It means that the costume design shall have careful thought. We built a glass nest room in the tree.

At the nest, there is growing the seed—costume design show. The designer taking it as the starting point, find out the most suitable point and elements of the cultural center and let them run through the vertical or horizontal direction of the space. The influence and creativity of the costume and culture center are merged into it.

The multi-functional performance room at third floor, taking the outline of Wanhua District Street as the inspiration, we change the street shape into the hanging rail on the ceiling. Through the multi-function and flexible design concept, the space can be changed in accordance with the actual needs to fully display the audio-visual functions. It has become the maximum efficiency usage of the space.

For selecting color, the first floor is white, showing the purity of space, simple and beautiful, different but inclusive. The work space of the second floor is calm and generous, creating a highest quality space for people.

咏真接待中心

项目地点：台湾台北市
设计单位：潘冠荣空间设计+筑筑国际
设计师：潘冠荣、陈欣
建筑面积：908 m²
主要材料：旧有建物、石头漆、回收桧木、白马装置
摄影师：李国民

白色斜顶烟囱小屋

斜顶小屋纯白质净的色调氛围，与大片草地坡地和羊肠小径交揉覆叠，欲引到访者置身宛如欧洲乡舍的绮思想象。908 m²的建筑基地和周边公寓绿波巧妙结合，形成更加广阔的花园前庭，刻意放大凸显自然场域，将接待中心本体浓缩成近99 m²的小型建筑基地置于后端，中间用圆形广场和蜿蜒小径拉出绝大部分的公共空间，本着一探究竟的好奇心，人们可能被夜晚发光发亮的屋舍、迷你钟楼或是其他装置艺术等细节吸引，让人体验感受空间地域性与物件彼此的连结关系，邀请更多访客、居民或是家庭好友到访，分享有别于都市节奏的乐活新象。

"UME FARME"（一个小农场）

拆建的过程中，设计师潜心保留原始老旧平房裸露的墙面结构，并以两匹白马装置艺术演绎古朴怀旧的马厩场景，在与L型弯角状接待中心的连接下，两者合而为一构成∏字型的半包覆空间，"UME FARME"（一个小农场）便是设计师为这个组合式建筑起的有趣小名。在墙上开启大小不一的正方形窗孔，其中的排列展演（其实是乐谱上拿掉五线谱记号的音符），以八分和四分音符组成C、B曲调，谱出即兴轻巧的小调乐章。另外别具心裁的部分是室内挑高2 m的桧木屋顶斜梁，使用从日本旧式宿舍拆建得到回收的桧木，将木材再次翻新打造，其中的接榫缺口和旧木质感比起全新木桩更具风味，亦可接续挪用在下一个空间，发挥更多绿色环保和永续精神。

Yongzhen Reception Center

White House with Slope Roof and Chimney

Taking the pure white color of the small room is to create a comfortable atmosphere. Coupled with the large grassland and small trail, they are attracting visitors to immerse a European rural homes like. 275 square meters base coupled with the surrounding apartment, they have formed a larger garden. The natural field that has been enlarged deliberately has concentrated this reception center body into a small building. The central place is set a round plaza and winding rails which have enlarged the public space. Upholding our curiosity, people may be attracted by the illuminated houses, mini clock tower or other art installation at night. They are allowing people to feel the connection relationship between the space and device. Invite more visitors, residents or family friends to visit and share the new life style of urban.

"UME FARME" (a small theater)

During the demolition process, the designer keeps the original old cottage wall structure and coupled with two white horses as the art device to interpret the simple and ancient horse hearse scene. Under the linkage of L-shaped corner like reception center, the two shapes are combined into a ⊓-shape semi-coated space. UME FARME (a small theater) is an interesting nickname called by designer for this building specially. There are opening many small square shapes on wall in different sizes. The arranged performances (in fact, they are the music notes that remove the music staff). They are consisting of eight and quarter notes to form C and B tunes, composing a light and smart music. In addition, the most attractive part is the wooden slope beam with 6 inches high, which is the recycled wood of Japan deconstruction room and is re-shaped. The joints and the quality seem more charming than the new ones. They can be reused by next space to play a more environment friendly role.

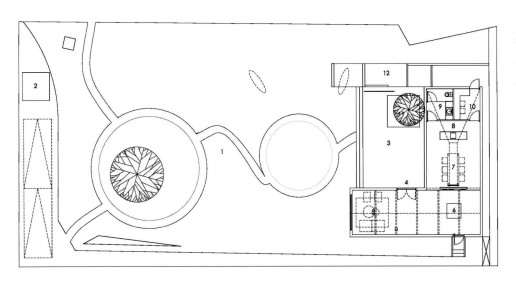

1	GARDEN	7	銷售區
2	警衛室	8	BAR
3	露台	9	衛生間
4	大門入口	10	OFFICE
5	沙發區	11	TREE
6	模型區	12	馬廄

Des imprimés de fleurs
pour des nuits en couleur

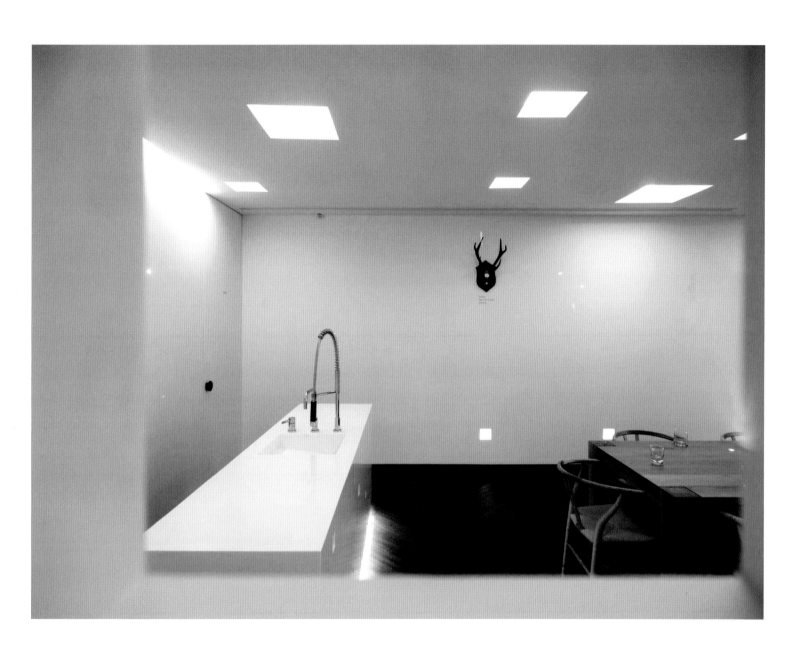

晶宴中和会馆

项目地点：新北市中和区中山路二段351号B1F
设计单位：大间空间设计有限公司
设计师：江俊浩
建筑面积：2645 m²
主要建材：石材、玻璃马赛克、贝壳板、银半反射玻璃、铁件、卡典西德、水晶板
项目时间：2010.05-2010.11

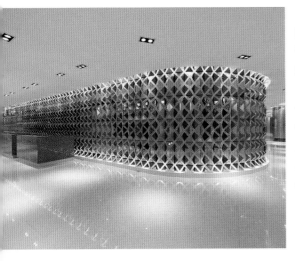

晶宴会馆以提供新型态且多样的概念式宴会空间立足婚宴市场，其企业品牌在市场上有着一定的知名度及领导地位。这个以完婚新主张及剧场型态为经营架构的企业品牌，企图在每间会馆创造出不同以往的空间美学设计，试图让消费者在短暂驻留的时间内感受到惊艳的空间渲染张力，进而传达其品牌在市场上的影响力。为了有别一般宴会空间的设计框架并符合惊艳的设计思维，设计师采取高度视觉感官接触的方式诠释其空间场域氛围，利用集体人流移动或伫立的影像轨迹映像于空间容器上，使其产生出戏剧魅惑的视觉影像画面，企图让人与空间环境达到奇遇式的相融，进而形塑空间的多样意象表情。

设计概念从表演事件的型态作为发想，设计师将其表演方式划分为被动式无感展演及主动式直接演出，利用空间容器转化成多重表情的演绎容器，让人置身于环境中时，会因不同的活动及不同的事件产生出不同面貌的互动关系，整体空间配置上主要划分三大区块——公共廊厅、宴会厅及后场空间。公共廊厅主要为人流汇集及过往的交通动线空间，利用其机能属性的特点，设计上采取将人形姿态影像被动地暂留于立面造型上，透过立体菱形镜的重复组合排列和色彩深浅变化，借此取得不完整且意象的画面，经由不同的角度及方位的

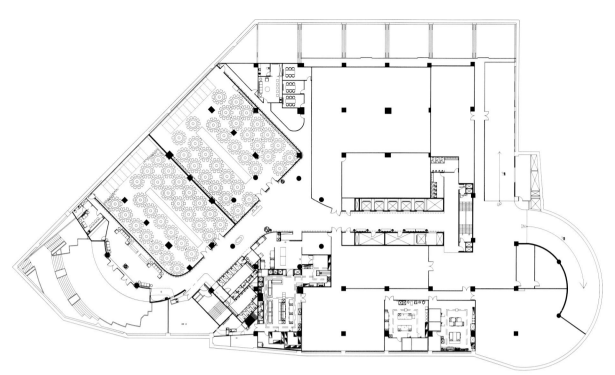

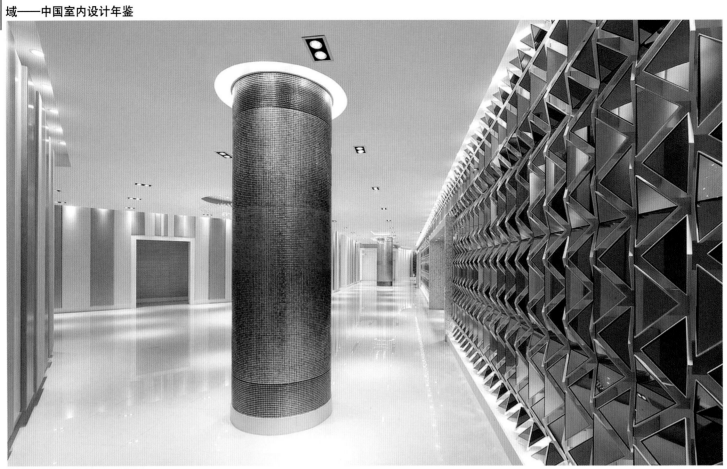

反射使其产生看与被看的无感表演故事。

为配合结婚新人即兴演出的各种桥段及短剧,宴会空间则采取直接表演的剧场概念呈现。用刚性交错的线条诠释的宴会厅,其色调采用较低沉色系铺陈空间氛围,炫丽多层次的灯光变化挥洒于天花板及壁面衬底的玻璃马赛克上使其幻化成极具变化的屏幕。菱形灯盒则随着表演剧情起落渐变光影凝聚众人目光。另一宴会厅则以柔曲的立面形体呈现,运用多层次的玻璃造型将光影层次染色其中,透过净透的表面材质将空间渲染出极具舞台张力的表演背景。将灯光影音系统整合于整体空间设计中,随着事件起落上演一幕幕人与空间的情境秀。

Zhonghe-Amazing Hall

The Amazing Club Banquet Hall offers wedding services and space. The brand name proclaims figure in this market and always surprises its innovative service.

It is well known for its amazing theatrical style wedding ceremonies and has won great praise from those who have participated in a wedding ceremony.

The interior of the designer branch is surprisingly colorful; it has the most flamboyant furnishings that offer a powerful domination of spatial attraction. The space is not a sequential experience from one space to the next but rather this is a space with a fully packed experience ranging from the visual to the acoustic.

The Design concept came from an idea that space is only realized through events.
There are two kinds of events; one is passive, the other active. The former offers each participant a surrounding spatial experience derived from the wedding ceremony. The latter invites each guest to participant in the wedding activity as if they are the ones who actually perform the shows.

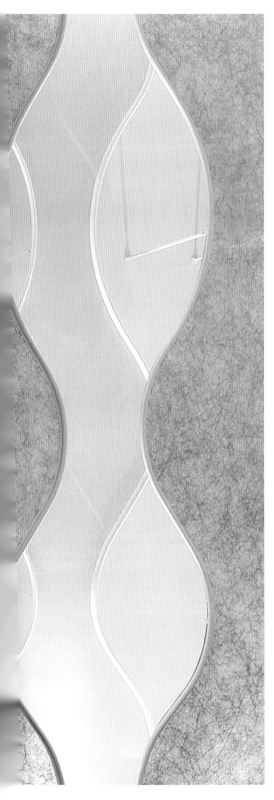

The entire commercial space is divided into three quarters: the lounge, a wedding hall and service space. The public is a place where every participant gathers and finds his or her friends. The space is like a stage awaiting everyone's chance to become a performer. A wall with a diamond patterned mirror is a unique design feature in this space as the wall reflects images of its surroundings and passing by human figures into de-constructed patterns. It seems that the phenomena of seeing and being seen has been distorted and shown in abstract form.

The banquet hall is a theater where the wedding ceremony is performed. The big hall is divided into two zones divided by the quality of lighting. One shows a more dramatic air with a lighting system controlled by computer programming. Light may come out from a ceiling panel clad with glass mosaic or laser-cut pattern board. Back-lit ceiling panels and flash lighting effects offer the space a dramatic ambience of pleasure.

The other side of the banquet hall provides a rather subtle air. The space has a strong visual impression in its curving wall while less colorful than the other hall. The dominate stage and its audio-acoustic system offers a strong visual focus to the stage and what is happening in the central position.

This is a large size project. In this project the designer was not only responsible for design work but also worked as a project manager who needed to meticulously coordinate many different jobs like the electronics, plumbing, and other support workers. But the most important thing was that the designer needed to make sure the quality of the construction was right and satisfied the client.

欧德旗舰店

项目地点：新北市林口区
设计单位：无有建筑设计 WooYo Archi
参与设计：刘冠宏、黄伟齐、刘艾玲、吴隆轩、洪钧莉、简伶倢、李重纬
建筑面积：**1652 m²**
主要材料：玻璃、铁件、石英砖、混凝土、木作
项目时间：**2010.09 -2011.05**
摄影师：李国民

　　此案位于地下室，因其采光限制，我们试着在偌大的空间中放入许多光盒，重新塑造一个新的室内，我们称之为屋中屋，屋与屋之间将定义为街道的意象。参观者优游于光盒与光盒中间，观赏着屋内的人与物，同时也被屋内的人所观看。光线被玻璃层层过滤而变得迷蒙，视线被折射与引导；所看到的世界，介乎于真实与想象之中。

　　此案因设计概念较为新颖，与传统商场相较，如何让业主接受为其最大挑战。有幸业主欧德陈董事长对于新事物有相当的包容度，并时常旅行国外，能接受西式的极简精神，方才使此案顺利完成。

Oude Flagship Store

This case is about 500 square meters in the basement. Due to the light limitation, we try to put various light boxed in the huge space to re-shape a new space, which we call it as tiny rooms in house. The space between rooms is called street. The visitors can walk among the boxes and enjoy the people and goods. The light has become dreamier after passing through the glasses. After fraction and guiding, the scene you saw is obscure and beauty, which is between the reality and imagination.

This case has a newer design concept. Compared with the traditional shopping malls, how to get the owners acceptance is the biggest challenge. Fortunately, the President Chen of Oude has a quite tolerant attitude. He travels abroad much and can accept this simple spirit of western style to make this case completed successfully.

早安清境

项目地点：新北市 汐止
设计单位：中怡设计
设计师：沈中怡
参与设计：杨佩珩、江易书
建筑面积：1楼 569 m²，2楼 684 m²
主要材料：钢构、玻璃、南方松、铁件、镜面不锈钢、黑镜、石材
摄影师：李国民

"早安清境"是国扬实业建案的接待会馆，基地靠近新台五路，近南港经贸园区，西南方面是几万平方米的森林保护区。设计师沈中怡便以建案最大优势——"与自然为邻"作为设计主轴。

由于基地本身一侧面临沿溪小路、一侧面临未来的计划道路（规划时，还没有开通），因此并没有正背面的区分，所有行经动线的视线都必然与接待中心的外观形成连接。于是，设计师将整栋建筑外观设计为蛋形，创造一种连续性环绕的视觉效果。同时采用"双层"的概念，里层空间由机能发展的量体堆栈而成，表层则加上一层蛋型实木格栅，呼应基地所处的自然环境因素。运用格栅作为表层的素材，很自然地让"光线"及"阴影"成为空间元素的一部分。

整个室内空间以素雅的木料质感呼应自然的室外景致并与白色基调空间相呼应，既是背景也是光影舞动的舞台。整体呈现如艺廊般的休闲舒适基调，室内动线经由刻意安排，最好的VIEW则配置给洽谈区，让参观者的互动能够与户外景致连结。

In The Morning Forest

The case of "Zao' an Chingjing" is the reception hall by Guoyang Construction close to No. 5 Xingtan Road, near Nankang Economic and Trade Park, with Wanping Forest Reserve to the southwest. Designer Shen Zhongyi has used "neighbor with nature" - the biggest advantage of this case as its design spindle.

Because the project land faces along the brook-side path in one side and the potential road according to planning in the future (when planning, it's not available) in another side, without distinction between front and back, all sights passing through generatrix are necessarily connected with the appearance of the reception center. Therefore, the designer has designed the entire architectural appearance in shape of egg to create the visual effect of a continuous surround. Meanwhile, "Two-tier" concept is adopted that the inner space is formed by the measure-body of functional development and the surface is coupled with a layer of egg-shaped solid wood grill to echo factors of the natural environment where the project land locates. The grill is used as surface material, so that it is natural that "light" and "shadow" become the part of spatial elements.

The wood texture of the entire interior space, simple but elegant, echoes natural outdoor view and reflects each other with space in white tone, as not only background but also stage with dancing lights and shadows. The overall tone shows as casual and comfortable as art galleries. The indoor generatrix has been deliberately arranged and the best VIEW is allocated to the negotiating area, so that the interaction of visitors can be connected with the outdoor scenery.

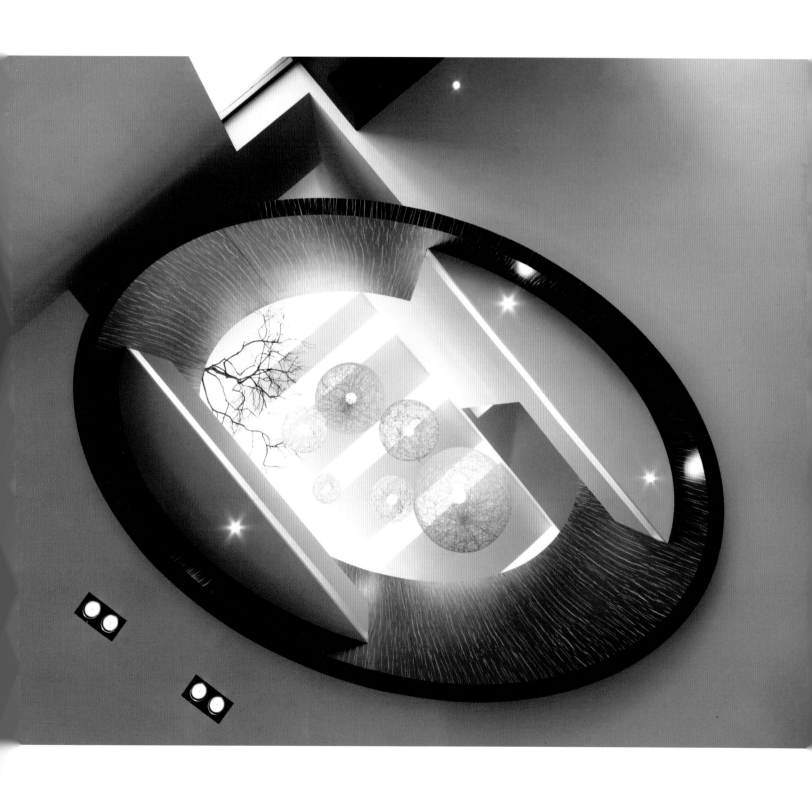

愿意

项目地点：台湾台北市
设计单位：近境制作
设计师：唐忠汉
建筑面积：1500 m²
主要材料：大理石、铁件、玻璃、地毯
摄影师：近境制作

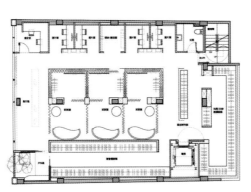
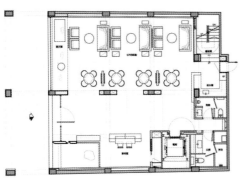
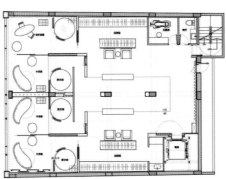

这个空间语汇里，白色场域中垂直与水平的黑色铁件装点了空间神圣的氛围。水平线空间轴线的延伸犹如红毯般一气呵成，银狐石墙比例切割与纯白进退墙面宛如婚纱裙摆的皱褶。逐层而上映入眼帘的是空间场域的转换，犹如女孩的蜕变。大面茶镜的反射，辉映了新婚的喜悦，白纱错落在白色量体与玻璃隔间中，犹如女孩珠宝盒里珍藏的幸福。借由艺术品的点缀丰富了视觉范围。窗外的绿意，石皮造景犹如来自大自然的祝福，创造了深浅有致、与众不同的视觉效果。

As we prepare to enter the space vocabulary, Field of vertical and horizontal white black iron. Decorating the sacred atmosphere of the space. Horizontal axis extending as though the space-like coherent red carpet. Silver Fox stone wall cutting the proportion of advance and retreat with the white walls. Like the folds of the skirt wedding. Layer while the eye is on the, Field conversion of space like a girl's transformation. Large seasoned millet mush reflect the mirror reflection of the joy of the wedding,The amount of scattered white body with white glass cubicle like a collection of well-being of girls jewelry box.Enriched by the art of embellishment by the visual range, Landscaping of the greenstone out the window like a skin blessing from nature.Creating depth has caused. Unique visual effects.

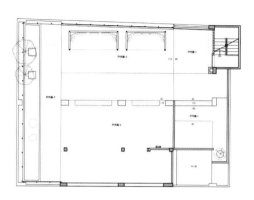

图书在版编目（ＣＩＰ）数据

　　域——中国室内设计年鉴.下/《中国室内设计年鉴》编委会编. -- 北京：中国林业出版社，2012.7
　　ISBN 978-7-5038-6666-1

　　Ⅰ.①域… Ⅱ.①室… Ⅲ.①室内装饰设计－作品集－中国－现代 Ⅳ.①TU238

中国版本图书馆CIP数据核字(2012)第148355号

本书编委会
编委：陈　彬　陈方晓　陈宪淳　蒋建宇　程绍正韬
　　　沈志忠　林伟明　陈德坚　Thomas Dariel　孔新民
编写：
柳素荣　史国丹　刘晶波　秦静思　尹星星　李　婧　孟　磊　杨　莹　谷艳凤　高　红
夏春雪　王　岩　蒋　曼　陈佳红　王　超　王　博　吕冬晶　娄　耸　李玲玲　周　爽
陈明阳　王　健　何　鹏　杨　旭　周　妍　黄　择　周　博　陆倩倩　陈　茁　王小燕
王亚楠　周　菲　于洪波　宋占生　韩　静　孙洪艳　张伶俐　张宁宁　雷树生　陶东颖
胡洪伟　崔力东　樊宏波　李　颖　李　蕊　王　焕　王　玲　刘　薇　王　健　刘景茹
策划：思联文化
采编：柳素荣

中国林业出版社·建筑与家居出版中心
责任编辑：纪　亮　李丝丝

出版：中国林业出版社
（100009 北京西城区德内大街刘海胡同7号）
网址：http://lycb.forestry.gov.cn/
邮箱：cfphz@public.bta.net.cn
电话：(010)8322 5283
发行：新华书店
印刷：北京利丰雅高长城印刷有限公司
版次：2012年9月第1版
印次：2012年9月第1次
开本：1/16
印张：21.5
字数：200千字
定价：338.00元 (USD 59.00)

购买本书凭密码赠送高清电子书
密码索取方式
QQ：179867195，E-mail：frontlinebook@126.com
法律顾问：北京华泰律师事务所 王海东律师 邮箱：prewang@163.com